SCIENCE, EVOLUTION AND CREATIONISM

THE EVOLVING HUMAN PRIMATE

AN EXPLORATION THROUGH THE NATURAL AND SOCIAL SCIENCES

SCIENCE, EVOLUTION AND CREATIONISM

Additional books in this series can be found on Nova's website under the Series tab.

Additional e-books in this series can be found on Nova's website under the e-book tab.

SCIENCE, EVOLUTION AND CREATIONISM

THE EVOLVING HUMAN PRIMATE

AN EXPLORATION THROUGH THE NATURAL AND SOCIAL SCIENCES

DONALD SHARPES

nova
science publishers
New York

Library of Congress Cataloging-in-Publication Data

Names: Sharpes, Donald K., author.
Title: The evolving human primate: an exploration through the natural and social sciences / Donald Sharpes (Emeritus College, Arizona State University, Senior Visiting Fellow, Cambridge University, Scottsdale, AZ, USA).
Description: Hauppauge, New York: Nova Science Publisher's, Inc., 2016. |
Series: Science, evolution and creationism | Includes bibliographical references and index.
Identifiers: LCCN 2015041023 (print) | LCCN 2016001203 (ebook) | ISBN 9781634840927 (hardcover) | ISBN 978-1-53612-242-8 (softcover) | ISBN 9781634840934 (ebook)
Subjects: LCSH: Human evolution--Philosophy. | Human evolution--Social aspects.
Classification: LCC GN281 .S37 2016 (print) | LCC GN281 (ebook) | DDC 599.93/8--dc23
LC record available at http://lccn.loc.gov/2015041023

Published by Nova Science Publishers, Inc. † New York

Dedicated to my executive editor, Marie, the joy of my life

CONTENTS

INTRODUCTION

This book is an exploration, through the lens of the principal social sciences, of how we live, and how our primate nature emerged from biology to define our true selves. I use the social sciences as a bridge to understanding our primate nature because they are the principal academic avenues for exploring our social behaviors and activities. The social sciences also provide useful theoretical and explanatory understandings of our conscious and unconscious behaviors. Our organic origins are situated in biology, but we clearly live in more complex social roles and groupings than other primates.

Although I seek to have the accumulated evidence of our biological heritage speak on its own behalf, and all the main social disciplines that address our adaptations have their collective voice, I also have personal views scattered throughout the facts on this synthesis of topics to illuminate our primate nature. The social sciences are piecemeal, arbitrary, academic pursuits—philosophy, economics, religious identity, sociology, political science, and anthropology, among others—that explore specific bits of how we organize our lives. (I omit education since its research typically includes other disciplines). So this study is interdisciplinary, a form of trend analysis, to unite disparate studies into a unified theme of our biological origins.

There is no collective, unifying theory about the social sciences. A doctrinaire view would be too simplistic, and anyway social problems are too complicated to conform to a simple formula. It's necessary to be open about all possibilities while maintaining a certain rigor about bringing the highest standards to the research craft. But I do think that anchoring the whole process on human origins in biology satisfies the quest for a theme without impairing other social investigations.

One of biology's key definitive words in adaptation, the process by which an organism adjusts its multiple sensory experiences to learn, survive, and acquire more control over its environment. A developing embryo, the immune system and the brain are examples of biological adaptation. My aim is to synthesize other academic pursuits into this significant paradigm of adaptability, to demonstrate how other studies of human social behavior can fold into this one primary life model and be relatively predictive.

John Holland provided such a unifying service in this groundbreaking 1975 book *Adaptation in Natural and Artificial Systems, An Introductory Analysis with Applications to Biology, Control and Artificial Intelligence*. He used biological adaptation to analyze economics, psychology, game theory and artificial intelligence. Although my study does not come close to his brilliant mathematical models, my goal of interdisciplinary integration, and the fact that we both begin with the complexity of evolutionary biology, is the same.

Similarly, Gillian Tett's 2015 book *The Silo Effect* reveals how cultural anthropology can illuminate the operations of corporations. She demonstrates how our inherited and ancestral social bonding methods of communication and collaboration are more creative and effective in business than the industrial models of bottom line economics and compartmentalization, to think outside the bureaucratic cubicle and departmental barriers, or silos.

Our evolutionary past and the influence of natural selection is credited rightly to Charles Darwin and Alfred Wallace who courageously challenged the prevailing orthodoxy of the 19th century by accumulating evidence of our animal history, and by telling the world where we really came from, and how much older that ancestry was than imagined. The theory of natural selection, the greatest contribution to the life sciences, provides the foundation for the accumulation of overwhelming scientific and confirmatory data, among them contributions from genetics, DNA, brain studies, comparative ethology, and even archaeology.

Thus, I have been influenced initially by reading Darwin's *Origin of the Species* (1859), *The Descent of Man, and Selection in Relation to Sex* (1871), and *The Expression of Emotions in Man and Animals* (1872), and Wallace's *Malay Archipelago, The Land of the Orang-Utan and the Bird of Paradise* (1869). Subsequent readings reinforced belief in natural selection, including E. O. Wilson's *Sociobiology* (1975), Richard Dawkins' *The Selfish Gene* (1976), J. Barkow, John Tooby and Leda Cosmides' *The Adapted Mind, Evolutionary Psychology and the Generation of Culture* (1992), and Steven Pinker's *The Blank Slate, The Modern Denial of Human Nature* (2002).

Other seminal studies that bolstered my persuasion were by William James, Bouchard, Gazzaniga, Cavalli-Sforza, Damasio, Paabo, Plomin, Scarr, and Mayr, among others. My own international research results over more than two decades of the genetic influences of adolescent self-concept, described in Chapter Six, strengthened my confidence that the genes we are born with have a predominant influence on our psychology and behavior.

Yes, we are members of a species known as *Homo sapiens* but we are not clones of each other, alike in some ways but palpably different in others. We also have group or sub-population differences as well as individual variability. Ethnic or racial population groups have more variability within them than they do between other ethnicities. All differential human classifications are controversial—certainly IQ—even though gender differences and athletic abilities are also predetermined by genes. Racial or ethnic differences initially were conditioned by geography and weather. Individual differences in physiognomy resemble cellular brain differences that result in personality distinctions, as I'll explore in Chapter Six. Other differences arise not only from birth but from experience and maturity, as older people generally tend to be wiser, less impulsive, and more practical.

I reject the notion of a human nature based on religion, as it has no trail of validated evidence to support its assumptions, and, at least in the Bible, unequivocally states that animals were created separately. Evidence from science will certainly not contain the indignant wrath and animosity of the religious right, who, to maintain their creationist or Intelligent Design beliefs, consistently oppose secular scientists and even science itself.

Religion also usually maintains that humans possess an immaterial substance, either mind or soul or both. Philosophy upheld these theological beliefs for centuries, and it was dangerous and often lethal to hold otherwise. But I believe we are not blank slates (John Locke). We do not have ghosts in our operating machines (Rene Descartes). And we are not noble savages (J.J. Rousseau). We are the highest order of the primates, inheriting all the characteristics of that group, and still evolving.

Even though I believe the nature of humans can be chiefly understood through biology, and that there are cultural, class, and sub-population differences, I reject the notion of racial superiority. I believe that there are broader social, aesthetic and common ends to human existence that are not exclusively rooted in the struggle for survival, or any other nihilistic, purposeless belief, such as the zenophobic and racist philosophy of Hitler governing Nazi Germany for over a decade, or the Islamic religious imperative of the ruling domination of a Caliphate.

We are the living products of millennia of evolutionary development, in conjunction with the spectacular wild of nature both on land and sea. Earth is our home, though many would like to escape it from time to time. Some believe that there are planets in the universe similar to ours that are capable of supporting human life. The odds of that are astronomical, like the universe itself. Besides, getting there might take more than a lifetime of frequent flyer miles in cramped quarters, never to return.

We have near-worship relations with celebrities, almost like members of a primate group who adhere faithfully to the strength of the alpha male or female. Young men bond with others to form gangs just like bands of wolves, feral dogs or outcast chimps, gang members adding rituals, tattoos and logos to symbolize their collective identity as they, like other mammals, engage in aggressive behaviors outside the tribal or clan membership. Then again, it is also like female primates of Park Avenue who strut their status with clothes and accessories associated with wealth.

We socialize with family, tribes and clans or extended family because we need group cooperation to survive—with the males to hunt as a team, for the women to gather and process food, to raise children and defend territory. We pretend our actions are human when they mimic the actions of other primates, or our more primitive ancestors very closely.

I assemble bits of available scientific knowledge, evolutionary research conclusions and logic to challenge past premises about who we are as humans. I propose that we are only slightly different species from other developed mammalian organisms, and that our description of "human nature" is an inadequate description of that difference. Humans do not have a separate kind of nature, although authors in philosophy, literature, history and religion have tried over the centuries to elevate our biological status beyond what it actually is. We are not just atoms and molecules and proteins, a regular part of the earth, but religious and social beings, bonding together to form all kinds of states, unions and relationships.

However, having admitted that the typical description of human nature description is passé, I admit that David Hume's monumental classic, *A Treatise of Human Nature* (1738-1740), is one of my exemplary models of the rejection of the religious and metaphysical accounts of human existence, to a radical departure based on sensations, perceptions and personal experiences, thus at least by the 18th century alluding to physiological foundations.

I've been a Jesuit seminarian in my youth, and in adulthood a federal government civil servant, and a university academic, with over half century of experience at multiple universities throughout the world, pursuing teaching and research in the social sciences, teacher education and humanities. Now I want to share my senior perspective, born of decades in the public sphere and a lifetime of study and research about the human social condition.

I am a member of a dwindling population born during the Great Depression, in the second year of the Franklin Roosevelt Administration, when everyone I knew was poor. As a toddler in my first house, next to railroad tracks, we had no indoor plumbing and no electricity. We used kerosene lamps for light at night, and an outhouse as a toilet about 25 feet from the house in the backyard. Our first telephone was a four-party line that all four families had to answer to discover who the call was for, and all stayed on the line listening to the communication of another family.

I am one of those who listened in my grandparents' kitchen to Roosevelt's fireside chats, and to one or two of Hitler's speeches broadcast with later translations and commentary on the radio. As a child I lived through and experienced the all-consuming World War II, when nearly 16 million Americans, including nearly every adult male, fought overseas in World War II. It was a time when America had to endure personal and collective sacrifices.

The point is not about my personal past but how in my lifetime conditions for living have changed the world in which we navigate. The old adage, adapt or die, is still true for actual primate survival as well as for accommodation to cultural changes, such as the latest technological gadgets.

My youth was also an era when people had a great amount of control over their lives and had to fend for themselves, to survive even on a modest scale, as my family did. That cultural world of modest provisions and cottage industries has disappeared and been replaced by corporate and institutional life where personal control is extremely limited. Today, we are governed by organizations that regulate our food production, finances, housing, transportation, health and entertainment. There appears to be little wiggle room for how we can manage the major conditions and events in our lives. Conservatives have a point about limiting the role of the federal government in private lives, loudly proclaiming the importance of individual liberty, but underestimating the role of corporations in controlling whole industries—processed and packaged foods, pharmaceuticals, corporate health, banking and insurance, and vehicles, to name some—that regulate and manage how we live, often at the expense of public safety.

Few can any longer grow their own food, build or remodel their own homes, gather their own fuel, or concoct their own simple entertainments and pleasures. All these activities my family did in my youth, in the manner of traditional societies. But such normal survival activities are now in the hands of corporate entities that charge a pretty penny for their services. We have been ambushed and imprisoned in a globalized and institutionalized world created by others. Inventions and technologies have indeed improved our lives, even in life-saving ways. But I conclude that such transformations have also come at the expense of personal liberties, the loss of choices and social values, and by increased costs.

I want to use the full spectrum of the social sciences, together with personal anecdotes, to investigate humanity, of who we are. This study does not confine itself to any particular topic, or a single academic discipline. Instead, it puts forward a synthesis of history and contemporary life, linked together in a unifying and integrated narrative of human existence. It builds upon my previous social science books, *The Evolution of the Social Sciences, Advanced Educational Foundations for Teachers, The History, Philosophy and Culture of Schools, and Advanced Psychology for Teachers*.

The imprecision of the social sciences has not deterred me from studying how humans interact, how aberrant their behaviors can be, and how ignorant, misguided, and frequently violent humanity often is. I have concluded that, at this stage in my career, using the totality of experiences and the fullness of all the social sciences is the most appropriate model for investigating social problems, especially one so intriguing as the status of the human condition.

Giambattista Vico (1667-1744) is the inventor of the social sciences and the study of culture. He was the originator of a new methodology in the social sciences by recommending that we enter into the mentality of history and the myths and languages of other cultures. He believed that culture gave meaning to how people actually lived, an understanding of motives, fears, loves and hates, hopes and ambitions. He rejected the methodology of mathematics as an inflexible tool for studying the complexity and variability of human affairs.

Social science inquiries are often inexact, making conclusions from such investigations imperfect. In the rush to publish, many investigators, especially in psychology, have used sexy social topics, imprecise methods, and over-generalized conclusions. In haste and inexactness, many such publications have been withdrawn or retracted because, when re-tested, results were found to be not the original outcomes, or with the same level of statistical significance, researchers claimed.

But that does not mean all results are irrelevant. Conclusions drawn from pervasive research evidence—at a time when slogans are seductive and mask more significant variables—can inform democratic debate, expose presuppositions, biases, expose fraudulent ideas, and lead to new insights. My goal here is grounded on my research over five decades in religion, science, history, psychology, politics, and culture.

Nor am I naïve about the contentions, embroilments and squabbling embedded in social science discussions. Name any emotion, gender, human ability, trait or behavior and you will find argument and opposition, as many theorists base their understanding of human activity strictly on the concepts of society, culture, religion or politics and not on biology. I do not intend to describe and sort out such controversies but rather to use the broad outlines of a particular social science, and the work of a few key proponents, to flesh out the evolutionary tendencies in the human primate.

In order to synthesize human social behavior, however, I must describe the origin of life itself, the physical evidence available in the professional literature that points to our evolutionary past. One way I describe myself is as an evolutionary psychologist, a social scientist and former teacher educator who believes in the biological progression of all living things. Humans, like all living things, have evolved over millions of years from humble organisms into the primates we are today.

All organic life, animal and vegetative, complex in intricacy and innumerable in form, has its origins in a common ancestor. Their existence has been patiently developed over billions of years and undergone multiple mutations to achieve what we now call separate species, though that term now has more ambiguity as a category for understanding life forms. Humans are at the apex of this long and formidable series of molecular adaptations, and it is their story that I will expound upon in this book.

The descriptive history I propose is based on the principle that natural selection is the exclusive explanation for the age of evolutionary development of humanity, and, barring any major catastrophe that decimated the dinosaurs, will continue to evolve into new life forms that either expand the meaning of a human primate, or into novel and as yet unrealized form of human life.

For example, one of the more visible and protracted behaviors of all organisms is aggression. It is natural to fight to protect the individual self from attempts at assaults, simple mastication, or worse, total absorption into a larger and more dangerous prey. Aggression to individual or group threats is present in fish, animals and humans. I will continually refer to this universal and genetically inherited trait because aggression, as both a personal and collective tendency to protect from harm—whether family, clan, territory, ethnicity or fanatical religious identification—defines the human primate in multiple ways.

We tend to consider acts of aggression as culturally-derived norms associated with rituals. We fight for the flag, a piece of earth we consider sacred, an identified territory to keep out those not a part of our group identity, in defense of the group's females or the brood, or our particular religion. The association of our defense capabilities is correlated with these reactive postures, so it is natural to assume they are related. In fact, genes drive our aggressive tendencies from real or perceived threats, as it does all marine and land animals in individual self-defense or in defense of the group. All gregarious animal and primate species will gather together in unity to repel aggressors.

I will use the reality of aggressive tendencies in the human primate to illustrate how similar it is with other animal species and to demonstrate its sociological significance. I don't include separate chapters on sociology or anthropology but weave the evolutionary traits of social bonding and group behavior throughout the other relevant social sciences to show how pervasive group cohesion is and how it is, or is not, repressed and conflicted as circumstances change.

The journey into personal memory is also a symbol and metaphor of the very long journey of the human primate. When we reach back to recall events or individuals from our past we usually have hazy recollections that may not conform to the reality as it was. Only the rare few, like Marcel Proust in *Remembrance of Things Past*, can describe in telling detail in one of the world's most literary writings his personal inner memory journeys with such descriptive power.

It's a common experience as we age to more vividly remember childhood remembrances than to recall yesterday's lunch. It is even more difficult to imagine the very distant past of the origin of mammalian or primate life. Our projections backward in time are as mentally speculative as mathematics and theoretical physics. But the scientific and archaeological evidence is real, valid and compelling, so our belief in natural selection is confirmed.

This inquiry has been a long exploratory journey for me and I'd like to summarize part of its intellectual history for you too, interlaced with a few personal comments and anecdotes as we examine, as on a scenic tour, the many ways of looking at our ancestral history. I hope you find that the adventure of continued learning about yourself and all of us primates will be as exciting as it has been for me over a lifetime of a fulfilling career.

Professor Donald Sharpes
Emeritus College, Arizona State University
Senior Visiting Fellow, Cambridge University
Arizona, US
Email: dsharpes@asu.edu

Chapter 1

LIFE ORIGINS

Observe how system into system runs,
What other planets circle other suns . . .
(Alexander Pope, An Essay on Man, 1734)

STAR STRUCK

There are some bodies and events in the universe that truly are awesome and beyond understanding. Stargazing can literally take you out of yourself and force you to feel the insignificance of the person you really are, and that your personal life in the expansiveness of the universe, and on a medium-sized planet in the solar system, has a transitory and even futile existence. You cannot feel arrogant stargazing. You can only experience humility in looking at the ever-expanding universe and staring into that stubbornly, persistent illusion Einstein called time.

Even today I keep taped to my refrigerator a Hubble photo of *Eta Carinae*, a massive star 7,500 light years away. I look at it daily just to keep everything in perspective. One day, like all stars it will explode, though we may not be around to witness its fiery extinction. In 1843 it flared up such that it outshone every star in the night sky.

My answer as to why the universe exists at all is that we have to wait for further evolutionary human development with more intelligent geniuses than those existing now to propose new meanings. At the moment, neither the God hypothesis nor the science propositions for creation from nothing into something are adequate explanations, even if we acknowledge there was an Alpha point, the so-called Big Bang, to creation of the universe.[1] There are billions of fiery orbs out there but, apart from electromagnetic and gravitational fields, no empirical evidence of what they are moving around in, or where they are going.

To better understand light years, the ultimate and standard measure used to calculate distances in the universe, let's take a minute to recalculate in planet years when the light from this one star, *Eta Carinae*, left its mass to make its journey to be visible on earth. And keep in mind that, as far as we know, nothing travels faster than light at 186,000 miles a second. What we see today is the 7,500 year-old light of *Eta Carinae*, a light that actually began in the Copper Age when Ice Man was still alive and hiking in the Alps to cross the mountains to find flint, at a time when farming slowly replaced hunting as a food source in Europe. It would be 2,000 years before writing was invented. Now that's a formidable time and space perspective, indicative of the vast expanse we call the universe and, yes, how insignificant even the human primates at the time were, fresh out of the Neolithic Age.

When stars do explode we might not even be here to witness the aftershocks. Yet at 2:12 AM on the early morning of March 19, 2008, anyone still awake and staring at the night sky would have been able to see a bright light glowing for about ten minutes. It was an exploding star about 40 times the size of our sun that occurred 7.5 billion years ago.

Time and distances like this are not, and never will be, within human experience. That's why it's astonishing to imagine this wondrous perspective. There are too many imponderables and the universe is one of them. We have to abstract from our normal understanding of how things work. But it is how this universe works. It is all about matter and energy taking place in time and space, the mental constructs we (with a big help from Einstein and subsequent scientists) use to describe the universe of physics, even though the reality may be something quite different.

Light travels at 5.9 trillion (that's 5.9 with twelve zeros) miles in a year. So this particular event actually happened before the formation of the earth. Thus it's possible to witness within the universe of matter and time the occurrence of an event, not only before humans existed, but prior to organic and geological life and the planet itself. I don't know exactly what the mind does when it gets boggled, but in addition to listening to Beethoven, Mozart or Mendelssohn, this might be a good example. Try putting that singular event in the context of commuter traffic, cafeteria selections and the price of gas and apples, or, for that matter, what it means to be human.

Amazingly all our individual atoms were once a part of stars, as my colleague and theoretical physicist Lawrence Krauss has explained. Exploding stars cooled to form planets and whirling orbs. Metals and rocks diverged into chemical compounds that formed gases and then liquids. Amino acids evolved into organisms. Iron, calcium, phosphorous, a critical component of DNA, and oxygen are some of our bodily ingredients that are left over from exploding stars. The cosmos may be distant reminders of the vastness of our universe and its longevity, but remnants of its particles live within us daily. We may believe that alien creatures inhabit outer space, but cosmic origins lie inside us. The organic building blocks of all life on earth were constructed somewhere else in the universe. Suddenly, the stars don't seem so remote since we were once a part of their matter. Even the New Horizons spacecraft took nine and a half years to reach the dwarf planet Pluto at the farthest reaches of our solar system.

Living organisms evolved from complex combinations of cellular structures. If you believe that the spatial universe is infinite and that everything consists of portions of all other kinds of things, then you believe what Anaxagoras did in the 6th century BCE. The universe is the science and the philosophy of us, that's where we came from, and that's who we really are—stray bits of the early universe. When we die we return to the earth. And when the earth dies, it, and all the atoms that once composed us, return to the stars where the atoms originated.

The sublime poet and philosopher, Rabindranath Tagore, winner of the Nobel Prize in Literature in 1913, wrote from his personal experiences of astral origins and in his inimitable style.

Light, as the radiant energy of creation, started the ring-dance of atoms in a diminutive sky, and also the dance of the stars in the vast, lonely theater of time and space. The planets came out of their bath of fire and basked in the sun for ages.
(R. Tagore, *The Religion of Man*, 1930)

We pass through time as light does in its perpetual travel throughout the universe, bending only slightly, or curvilinearly because of gravity around an object like a star, but moving forward as if nothing will stop us, hoping to land on our feet somewhere pleasant.

Nature is random and chaotic. Tornadoes, earthquakes, volcanic eruptions, tectonic plate movements, the passing of comets whizzing by the earth and star explosions are all random. There is no grand design to nature's activities. Why then do we sometimes believe that events that occur to humans can be controlled by propitiations to divinities, referrals to astrological charts, or the tea leaves and crystal balls of psychics?

I am one of the fortunate few who have rafted down the last 55 miles of the Colorado River in the Grand Canyon before it descends into Lake Mead. Like all who have experienced the Grand Canyon from the north and south rim sides, or from floating down the Colorado River, I can attest to its daunting experience and majesty, a place of poetry and exquisite natural beauty unrivaled in the world. The steep canyon walls house 20 distinct layers of stone from the top of the rim to the river one mile below. These rocks contain 1.7 billion years of geologic history, or about a third of the life of the earth itself.[2]

Is there life similar to that on earth elsewhere in the universe?[3] The search for extraterrestrial life (SETI), a specialty of astral-biologists, is a combination of star-gazing to discover other planets revolving around other stars, and microbe analysis here on earth to see how small life forms that exist in extreme conditions. Thus far, the search has proved to be fruitless, even though Steven Hawking, the guru of theoretical astrophysics, believes that some kind of life exists somewhere besides earth. To date, nearly 500 planets have been discovered outside our solar system. None of the planets within the solar system have been found to have organic life. Perhaps life forms as we know them need a few more million years to evolve into our peculiar primate state.

LIFE FORMS

Despite all our fantasies and pretensions, we always have been and will remain a biological species tied to this particular biological world. Millions of years of evolution are indelibly encoded in our genome.[4]

This is astral story is about us. But it begins much earlier, prior to our arrival on earth. I think it is important to review what happened before the appearance of earthly life. We are indeed the rocks of stars, but there is much more that preceded our long and delayed evolution. Here is a brief description of what we know about the history of the planet we call home.

The earth was formed about 4.5 billion years ago. Gases like hydrogen cyanide mixed with metals about a billion years later, and the resulting chemicals combined in a variety of ways to form lipids, nucleotides and amino acids, the building blocks of organic life and the components of a living cell. Ribonucleic acid, or RNA, is a nucleotide, a string of sugar (ribose) at one end and a phosphate group at the other end, and has the ability to duplicate itself. Scientists now assume that RNA preceded DNA in origin.

A billion years later the first life forms emerged, simple cells known as prokaryotic bacteria, the only life that existed on earth. But two billion years later, by 1.5 billion years ago, cells known as eukaryotes emerged. Eukaryotes evolved from prokaryotes cells like the E.coli bacteria that have no nucleus.

Eukaryotes have the ability to switch their genes on and off and can control how they made proteins, the essential building blocks of all life. Eukaryotic organisms have multiple cells with a nucleus, membranes, chromosomes and sexual reproductive cycles. They are the original ancestors of all animals, plants and fungi. Eukaryotic cells further diverged into numerous biological classifications with increasing complexity. By the time of the Cambrian era, about half a billion years ago, multi-cellular organisms swam everywhere in the primitive oceans.

Evidence of life forms dated to the Cambrian era is just a small part of the continuously unbroken life force. One such creature is not a worm, but the two-inch long *Pikaia*, a primitive spinal column (or chordate), and the world's first vertebrate animal. If *Pikaia* had not survived the Cambrian era, neither would robins, fish, deer or us, the human primate.[5]

Trying getting your mind to think back 500 million years ago when the earth teemed with a variety of sea creatures. Then consider imagining how one of these sea creatures, *Metaspriggina*, very slowly over several millions of years, evolved into eels, sharks, other kinds of fish and mammals.[6] *Metaspriggina* had vertebrate features, two large camera eyes, twin nasal sacs, a liver, gills, back muscles, and a primitive cranium. Already, only 50 million years after the beginning of the Cambrian era that began about 600 million years ago, ancient fish have developed anatomical features we would recognize because we know and have seen similar marine creatures.

Similarly, evolutionary processes at the microscopic level have occurred rather quickly as exemplified by the mutation of genes in some bacteria to become drug resistant to penicillin. In fact, totally resistant species of bacteria have evolved to become immune to the strongest doses of all antibiotics.[7]

All life is a marvellous expression of the diversity and complex organization of cells and the evolving structure of molecules. This conclusion is as mysterious as any creation story that describes the origin of life or human existence. We are composed of a mass of microorganisms. Everyone has at least 100 trillion bacterial microbes, outnumbering our cells by a factor of ten. Welcome to another awesomeness of nature, as awesome as a violin concerto played by Itzhak Perlman on a Stradivarius.

All the major subdivisions that would evolve into animals and mammals existed in the Cambrian era. Although the dates are rough conjectures, fishes appeared about 450 million years ago, reptiles 310 million years ago, and birds and mammals about 225 million years ago. The coelacanth fossil fish is dated to about 60 million years ago. Everyone thought it was extinct until two living ones were caught off the South African coast in the 20th century. The oldest undisputed fossil of a bird is the *archaeopteryx* that dates from 145 million years ago.

The theory of Punctuated Equilibrium, proposed by Stephen Gould, is one example of how evolution occurred.[8] Punctuated equilibrium speaks to the origin and deployment of species (phenotypes) in geological time. Once established, species remain relatively stable over long periods. It isn't clear how evolution proceeds within a species, as each form might remain relatively unchanged over long periods and then quickly adapt to new conditions.

Alternatively, a species like crocodiles might coast along over centuries with only minor changes.

TAXONOMIES: CLASSIFYING SPECIES

Species is not always the most accurate term, either for us human primates or for variations in the vegetative or animal world. A more appropriate term is *varietals* or *lineages*. Primates vary tremendously in shape, size and sometimes in facial and behavioral approximation to humans. One particular Japanese macaque or Snow Monkey looks like one of my former grandfathers, though granddad had less dramatic hair. Their eerie facial proximity to humans makes them so attractive and fascinating. Snow Monkeys cavort in a local hot spring and forested region near the town of Yudanaka, about two hours north of Tokyo, and farther north than any other non-human primate.

Lemurs, on the other hand, are much older in the evolutionary tree, on the earth's endangered species, and date from the Eocene era, about 55 million years ago. One lemur branch, known as *Coquerel's sifaka*, live largely in Madagascar and they have lost about 90% of their native habitat because of slash-and-burn agriculture by the natives. The phrase "There goes the neighborhood" in this case means lemur extinction.

Cro-Magnons and Neanderthals were also varietals of *Homo sapiens* approximately 150,000 to 200,000 years ago. Scientists like Ernest Mayr define species as the ability to interbreed and have young that can live on their own in the wild. DNA evidence supports the theory that Cro-Magnon peoples and Neanderthals once interbred and that everyone today has a small amount of Neanderthal DNA. We are not as unique as we think, just bits and pieces of previous organisms evolved over very long time periods.

Who we are, biologically, is a combination of 25,000 genes arranged on 23 pairs of chromosomes. In a nutshell, that is all of us. What makes the distinction between the word "species" so difficult to interpret is that not all chromosomes within an individual are exactly the same for all members of the species.

Variability within any species is the rule and not an exception, which is why mutations can occur. It is because of mutations that species will eventually disappear and new ones emerge in the long chain of evolutionary life. Paleontologists agree, following the work of the pioneering paleontologist David Raup, that though we may never know how or why species become extinct, that more species have become extinct than now exist. This should make us less sanguine about the ultimate survival of humans who, as we'll see in a subsequent chapter, are starting to slowly destroy the planet.

In the past, biologists came up with the term "subspecies" to denote variation in a species, as it did in 19th century anthropology to explain human racial differences, mistakenly as it turns out. But this too did not explain all the variation. This is why I prefer the term *varietal* to explain similarities and differences within a particular type of organism where all the members look somewhat alike. The slight differences in beaks among the finches Darwin found on the Galapagos Islands are examples of variability within a species.

TETRAPODS: GETTING A LEG UP

Until recently, a sparse fossil record obscured efforts to trace the evolutionary steps leading from fish backwards to land-based *tetrapods.* The present fossil record allows scientists to conclude that roughly 380 to 360 million years ago, in an era known as the Devonian, a group of mammalian precursors known as the *tetrapods* evolved, aquatic creatures about three feet long in the process of evolving into mammals. The closest living creature is the iguana. About 350 million years ago *tetrapods* existed all over the globe and have been found in China, Australia, Pennsylvania, Latvia and Greenland.

One archaeological specimen whose skeleton was discovered in Greenland is the *Ichthyostega*, a three-foot long, four-legged creature classified as a tetrapod, or a four-legged land vertebrate that lived 360 million years ago. Scientists believe that such ancestral fish developed legs and toes, not to walk on land, but to move in shallow waters. The earliest tetrapods were all believed to be aquatic, but that they came on land permanently to locate food sources and become mammals about 250 million years ago. *Ichthyostega's* closest contemporary relative is the elephant seal, and perhaps the whale, classified through DNA as a hoofed mammal, and possibly the hippopotamus.

They lived in shallow aquatic environments but took oxygen from the air because water at the time was oxygen-depleted. They were fish-like but yet distinct from fish. They had interlocking vertebrae, longer snouts and skulls separated from the shoulder. They had, like the fossil named *acanthostega* found in Greenland, front and hind limbs with an eight-digit foot and no fins, and lungs as well as gills. Though they had legs it is surmised that they were not sufficient for terrestrial use.[9]

The spectacular discovery in 2005 in the Canadian Arctic of a giant 375 million year-old-fish four to six feet long is considered a transition between aquatic and land animals.[10] This unique animal is another piece in the puzzle of links between water and land mammals. The fossil was named *Tikaalik roseae*, a Nunavut Indian translation meaning: "large shallow water fish." The find revealed how fish acquired characteristics of modern land animals in the Devonian period. In addition to forward fins, it had digit-like fingers, wrists, elbows and shoulders, a neck and ribs—all very un-fish-like. Fishes do not have necks. But since *Tiktaalik* had a neck it was able to swivel its head, a new evolutionary development. We think of fish fins like paddles, propelling the animal through water. But *Titaalik's* fins had the ability to flex and extend like arms and hands.[11]

This discovery adds to the accumulating evidence that evolutionary forces, with time spans of millions of years, can shape new and emerging organisms. The other amazing feature of this discovery is that paleontologists knew where to look for just such a creature—amid the geological sediments of a particular time period, the Devonian, on Ellesmere Island in Canada, a section of the globe that once was subtropical before the shifting of tectonic plates moved this part of the earth northwards.

Nature's unusual manifestations in mammals exist today. The tusk of the narwhal, for example, that can grow as long as nine feet, long valued and little appreciated. It was, like the pointer of the mythical unicorn, thought to be magical. Scientists have recently discovered that it is actually the whale's left tooth projected far forward that enables the mammal to measure small changes in water pressure and composition.[12] The narwhal's tusk with its ten million nerve endings, long one of nature's secrets, is one of nature's most remarkable appendages. It allows this unusual marine mammal the ability to explore and understand its environment with its unique tooth. Narwhals migrate to the Canadian arctic during the summer months where they nurture their young, but find that they are tracked by killer whales, their chief predator.

Even at the level of fruit flies, evolution has been found to occur quickly. Michael Rose, Professor of Evolutionary Biology at the University of California, Irvine, experimented with aging in the 1970s by letting fruit flies breed only at late stages in their short lives to test their survival and reproductive behavior. Over the next twelve generations fruit fly descendants had longer life spans and greater reproduction capabilities. His conclusion was that ageing is an evolutionary process. The same process in humans would take centuries. Medical interventions, like replacing cloned organs and cultured tissues, will probably prolong life even more before results from the evolutionary process.[13]

Some species become extinct through our carelessness and neglect. To learn how species and their varietals came about we have to descend into a microscopic world and study its schemes.

THE REALLY SMALL STUFF

My point is literally to return to the biological beginning, where we all begin, where the similarities exist that bind us together as a primate species, before exploring ourselves as behavioral and social primates. How nature duplicates itself has always fascinated me. We are the survivors of the original DNA, or deoxyribonucleic acid, and that fact is totally awesome, the greatest biological mystery of the universe. Genetic mutations cause genetic variations. Over millions of years, therefore, a wide variety of organic specimens have arisen to fill the earth with an exceeding wide pattern of life forms. To understand this process we have to imagine the incredible complexity in DNA, that twisted double helix pattern of all life, and then to imagine how that complexity can be recast, reformulated and recombined over vast amounts of time we can never personally experience.

All life forms are amazing in their complexity and multiple manifestations. But genes, the principal construction material of all life, determining eye color, blood type and personality, are the most amazing. Genes are the twisted, intertwining, double-helix forms that constitute the DNA molecule. Genes within the molecule make up a chain of chemical components: adenine, guanine, cytosine, and thymine, often abbreviated as A, G, C, and T.

If one of the strands ever breaks apart or separates by uncoiling or unwinding, each separated strand makes another set or template of itself. Genes are self-propagating life forms, the smallest elements in the chain of human shape and behavior. They literally govern how and what we know, and how we think or retain memories. Everyone is a human fossil carrying the genetic information of every ancestor and traits of the varietal. We are all a continuation and reproduction of the original DNA.

The age of DNA has recently been pushed back about 80 million years with the discovery and publication of a DNA molecule sequence from 120-135 million years ago. Scientists report that DNA has successfully been isolated from fossilized plants species and animals. The oldest DNA has been extracted from an extinct species of weevil found in amber in

Lebanon and dated to 120-135 million years ago, in an archaeological time period known as the early Cretaceous.[14]

DNA has also been extracted from insects deposited in amber as long ago as 40 million years, and they are in a better state of preservation than the Pharaohs. Paabo in the early 1990s reported on his collection of samples of DNA from ancient human Egyptian mummies. He discovered from analyzing one mummy who lived between 2,310 and 2,550 years ago that DNA molecules survive long after the death of an individual and can be partially reconstructed.[15]

The message of how life-forming information is encoded in the gene is one of nature's best-kept secrets. The message of replication is passed on to the cytoplasm of the cell where it is duly recorded.[16] The great project of biology, the Human Genome Project, has mapped the sequence of each of the genes in the DNA molecule. Infinitely small is where nature resides. The Human Genome Project was finally completed in 2006 after having sequenced 20,000 to 25,000 human genes, an amazingly technological and biological scientific breakthrough. What we cannot see with the eye determines who we are.

To give a current illustration of the power of genetic combination, genetic engineering has produced an agricultural revolution in the face of an ever-expanding human population. Insulin for diabetics was once extracted from pigs, but can now be reproduced through genetic engineering and its pure clone cultivated for human use. The hope among geneticists is that individuals carrying genes that lead to potentially life-threatening diseases, such as cystic fibrosis, Huntington's disease or cancer, can be located and treated. Genes control regional systems and brain centers crucial to an understanding of brain mechanisms, learning and even personality.[17]

We each have a few trillion cells that form us into an organic and personalized whole. We could spend an entire lifetime just counting each cell, as many as there are stars in the universe, as Carl Sagan constantly reminded us, and then the world would know more accurately just how many there are. Would having significantly more cells make us more human? Does having less cells make us less human?[18]

The world of microcosms also defines humans, as it does all organisms, often in unpleasant ways. The Ebola virus—really any virus—can cause lethal damage unless the body can resist it through the immune system. The Ebola virus eats through the membranes of major vessels in the body, like blood vessels that cause internal hemorrhaging everywhere, including the brain. The Ebola virus has no known cure, and no vaccine or drug can stop it.

Viruses and bacteria have no prejudices about species and, like squatters, decide to take up residence in any organism, whether invited in or not. It has only been since the latter part of the 19th century that we have known that germs are the causes of our diseases, and not spontaneously generated microbes as was once believed. Louis Pasteur's conclusive experiments were resisted at first. His persistence and experimental validation demonstrated the effects of the tiny, invisible enemies.

The pandemic of the Black Death, peaking from 1348 to 1351 that decimated from a third to a half of Europe's population, is perhaps the worst global example of how unseen organisms can kill. The disease was transmitted by a bacillus, a germ known as *Yersina Pestis*, through the air or from the bite of infected fleas or rats, those perennial rodent survivors. Bloodletting, bathing in vinegar and wearing garlic did not help. No one at the time understood how diseases spread.

A virus is not peculiar to humans and invades any organism to thrive. Porcine epidemic diarrhea, or PEDv, is also a deadly virus that kills pigs. It first appeared in Iowa in 2013 and within a year had killed over a 100,000 piglets intended for human consumption. As a result, the number of hogs slaughtered diminished and the price of pork and bacon rose more than twelve percent. The burial of so many pigs was said to have contaminated groundwater that people relied on for drinking. So a virus may not be susceptible or transmittable to humans, but the human cost can still be high in environmental damage and to the pocketbook.

Duplication: Getting New Ones

Each individual organism is the result of a selection of one sperm (except in the case of twins) penetrating the egg for fertilization. The single sperm is not necessarily the most "fit." Rather, it is the one chosen randomly as in a lottery. The ultimate goal of all sperm is to find an egg to fertilize, though few achieve that objective. In this sense each sperm is genetically determined. But the winner is not pre-determined and cannot be predicted. The same process that occurs in individuals occurs in other populations, or varietals, because of attributes favorable to survival.

But in addition to the mutations in DNA over time, consider factoring in chemical changes, even within a body such as yours or mine, over shorter periods of time, to understand how there is no static state in living things. The body's chemistry is constantly mutating and altering the metabolism. You are recognizable over the years, but your face does change because all the body's molecules change, decay, get replaced, and eventually slow down as you age.

Regarding body changes, I am reminded of Franz Kafka's bold fiction of 1915 entitled *The Metamorphosis*, a meaning that has the evolutionary distinction of a transformation from hatching to independent life. But Kafka brilliantly reverses this normal organic procedure and has his protagonist waken one morning to find that he had been transformed into a giant insect, a kind of reverse of the metamorphosis found in nature.

All organisms fall into many categories. For survival purposes the two main categories are prey and predator. Predator organisms eat other organisms. Prey organisms have many means to avoid this by flight while they search for their own food source. Lappet moths, when they emerge at the beginning of the dry season in Africa, have coloration patterns on their wings that look just like a fall leaf. Cryptic Mantis and Preying Mantis resemble tree twigs and often appear with the same color as the tree bark.

Both predators and prey have to use cunning as prey to avoid either becoming food for another organism, or as predator to fool its source of food into believing it is not a predator. In other words, the plan for survival is deceit—also a common human behavior—either through camouflage, or hiding, or menacing behavior. Deceit is a typical tactic used in warfare. In a similar fashion, all organisms have to possess persuasive devices to convince mates that they are the right one to produce offspring with.

The individual organism, like any of us, when it first enters the world on its own—though not yet on its own terms—knows nothing. The cells, however, do know something because they carry genetic knowledge about how to act around other cells and how to combine to create larger operating cellular functions. They also carry information that they share with other cells, and this blended knowledge can create a kidney or other complete bodily organ which links with all other organs to make a person. A person's life is the life of a cell magnified and extended.[19]

I am not an archaeological anthropologist and do not intend to revisit all the scientific names of our presumed primate ancestors, or to list all the Latin-named categories of differing lineages of primates that preceded us in the tree of life, or to describe the geographic dispersion of *homo sapiens*. But closer in time, and biologically as a primate species, everyone has as many as 5,000 ancestors just from the last 300 years.[20] The reproductive organs continue to supply the earth with additional primates like us.

Even vegetables have recognizable primate features. Have you ever wondered why radishes have tails, corn has whiskers, and most vegetables and fruits have skins?

But if our evolutionary past is so awesome, the ability to design, construct and install new organs is just as amazing. Cloning the whole human organism is now available but ethically contentious. In 1996 Dolly the sheep appeared, the result of the cloning of cell manipulation, to create an animal without sexual intercourse. Horses and cattle have since been cloned, revealing that any cell can be used to create an entire organism. Cloning of cells and DNA, often known as genetic engineering, is commonly used to create hybrid fruits and vegetables, like the hybrid cross between a kiwi and a tomato first created in 1973. Updating Protagoras just a bit: man is not the measure of all things. DNA is.

But soon most individual organs will be available too. Cranial plastic implants were first installed in a patient in 2013 to reconstruct a large part of the skull. A new mandible was placed for a woman's lower jaw in 2012 from titanium powder and a biocompatible ceramic. Toward mid-century, according to researchers, bioengineered lungs, heart valves and bone tissue will be available.[21]

Ida, Our First Known Primate and Us

Ida, at 47 million years old, is the most complete primate ever uncovered. She was found in the Messel Pit in Germany, an extraordinary trove of creatures from in an epoch known as the Eocene, covering approximately 56 to 34 million years ago.[22] She was found in 1982 by an amateur archaeologist who preserved her on a shelf for the next 25 years, until the fossil was spotted in 2006 at a fossil fair in Hamburg by a Norwegian paleontologist.

She is classified as a primate, the same as lemurs, monkeys, apes, gorillas, chimps, and us, and they varied in height and weight. There are unusual sizes of similar species like dogs, where dachshunds are like the mouse lemurs of primates to the Great Danes of the dog world. But all primates have similar features that include opposable thumbs, two pectoral breasts, and forward-pointing eyes. But Ida is the most perfectly preserved skeleton found in any age.

Ida measured 23 inches long. Her legs were longer than her arms indicating that she was a climber and leaper, living in trees. She had large eyes as lemurs do today, and analysis of her teeth showed she was largely a fruit and leaf eater. She had nails not claws. The Norwegian team that acquired her established that her diet, lifestyle, and size were similar to the Eastern wooly lemur living today on the island of Madagascar.

The stupefying experience of knowing about Ida is not just the taxonomy, or what her ancestors could have been, or even what descendants are similar to her today. What is astounding is her age. At 47 million years ago she looks amazing like creatures that exist today. At that age her species could very well be the type of organism that modern primates evolved from. The exact map of the evolutionary trail of primates may never be known. But that doesn't stop scientists from speculating about who is related to whom, and from which certain species are descended from which other species.

Physicians spend most of their professional lives studying the human anatomy and its variety of organisms and development. Imagine over how many millennia it must have taken for evolution to achieve the present status of the human primate. We already know that at least 47 million years ago lemur-like primates had all their parts and internal organisms in order. The next time someone asks you what famous person you might have as an ancestor, say you are a direct descendant of the first amoeba. There's a conversational stopper . . . or opener, depending on your audience.

ORANGUTANS: CLOSE PRIMATE RELATIVES

If your trudge through low swamps and through thick undergrowth and high forest canopies in Aceh province in northern Sumatra you may be lucky enough to hear a swooshing sound. That sound will alert you to trees swaying back and forth, sounding like swoosh, caused by orangutans leaping through the trees. These "people of the forest," as the native Dyaks of Borneo, and Malaysian and Indonesian natives call them, are also known as the red apes, with blazing red hair, the mysterious primates only found in Southeast Asia, on Sumatra and Borneo, and not at all in Africa.

They are surprisingly human-like in appearance. The large males weigh about 200 pounds, and orangutans live to be about 50-60 years old. Like humans they have a long childhood and constantly bond throughout childhood with their mothers who only produce a child once every few years. They are on the world's most endangered species, with only about 7,000 of them left in the wild, and 85% of them in this Sumatran province. The forest canopy provides fruit and seeds for consumption, and they sleep in tree nests to hide from predators like tigers and pythons at night.

The economy of the indigenous natives of Aceh province is built around palm oil. Planting palm oil—a product used in most cooking, cleaning agents, soaps and waxes—means that the natives have to set fire to the forest trees and undergrowth to clear the land, thus destroying the orangutan habitat. The task of the world community is to protect these and other endangered primate species, our very distant, but obviously close primate ancestors and relatives.

Alfred Russel Wallace (1823-1913), co-discoverer of the theory of natural selection and a naturalist, spent 12 years in the field collecting examples of natural life in both South America and Southeast Asia, nearly three times as much fieldwork time as Darwin. He lived for four years in South America and eight in Southeast Asia. He earned his living by collecting and forwarding back to England examples of natural life, including exotic birds and butterflies . . . and orangutans. He hunted these exotic and human-like primates to send skeletons to British museums like Derby and to private collectors. He mailed back to England over 125,000 specimens of all manner of natural life—insects, beetles, exotic birds like the multiple varieties of the Bird of Paradise, butterflies, animal skeletons and skins—after having acquired them in extensive travels through the region from Malaya to Java, Bali and Timor, the Celebes, the Moluccas, and Papua New Guinea.

Wallace had come to a similar conclusion about the evolution of natural life independently, a debt that Darwin acknowledged when both had abstracts of their works read at the Linnean Society in 1858. Wallace was still collecting biological samples in Southeast Asia when Darwin published *The Origin of the Species* in 1859. As a curious traveler he was also a budding anthropologist and collected linguistic vocabularies of scores of the various tribal groups he encountered.

On one of his first orangutan hunts, he killed a female and found that its infant had fallen face first in the mud. He cleaned it, took it to his residence, and it stayed with him as a pet for three months while he fed it before it died of an unknown illness. Wallace observed that it acted in so many ways exactly as a human infant acted, with whimpers when it wanted attention, with facial expressions of food it liked and didn't like, of playfulness with other monkeys, of cuddling next to Wallace as if it would an orangutan parent. He would have liked to have raised it and brought it back to England with him as an adult.[23]

Wallace also had a keen insight into the allied nature of primates as having more ancestral origins, as he notes in his classic study, *The Malay Archipelago* based on his explorations there.

> *When we consider, further, that almost all other animals have in earlier stages been represented by allied yet distinct forms . . . all different from any now existing but intimately allied to them—we have every reason to believe that the Orang-utan, the Chimpanzee, the Gorilla, have also had their forerunners . . .*[24]

Scientists agree that we don't know everything we'd like to know about our humanoid origins. But in recent years they have added more branches to the ancestral tree, discovering new fossil forms in different corners of the world that resemble us, some of which became extinct.[25] That branch of the tree that emerged as us, *homo sapiens*, the human primate, took a long time to get everything in place that looks like us now, most of it— including the long legs, the enlarged femur, the low shoulders, and increased cranial capacity—took about two million years. That is not a long time in the history of the universe or the earth, but does show how long it takes for adjustments to occur in our evolutionary development.[26]

Charles Darwin proposed that all animals have inherent traits that can be duplicated. Successful traits, he wrote, get passed on to the next generation.[27] At the time, Darwin did not understand how these traits were passed on. Gregor Mendel (1822-84), a cleric recognized as the father of modern genetics, gained notoriety posthumously with his research on peas that demonstrated how what he called traits are inherited in subsequent generations. At the time, genes had not yet been discovered. What is known now is that genes control everything from peas to humans.

As primates go, the yellow baboons occupying the Luangwa Valley along the Luangwa River in eastern Zambia, exhibit similar social traits as humans. There is always a dominant Alpha male, or Troop commander, supervising the group. Like urban human primate gangs, rival males who attempt to enter the territory are quickly driven out, often by juvenile males anxious to demonstrate their fighting ability, sped and agility. The troop commander has rights to all the group's females so there is never a question of paternity. The females want the genes of the toughest sheriff in town so the offspring will be strong too.

There is also a pecking order, a hierarchy among the females, with a recognized dominant female, usually one of the mothers. Rank in the group among the females establishes certain privileges, including the approval of the chief male. Strange as it may seem, this Alpha female domination exists among chicken hens. The Alpha female hen was always the first to eat in the chow line followed by the hierarchical pecking order, literally, among the hens. The first among equals among the hens pecked the others in the eating line but was never pecked in return.[28] The hens learned their place in the dominance of the Alpha female chicken. Hierarchical dominance exists in lower order animals and even in hummingbirds.

There is a great deal of social interaction among the baboon tribe or troop, much of it around grooming and picking mites from the fur, adding protein to the diet of fruit from the ebony trees, but also in nurturing the young. A grieving mother will carry her dead infant around with her for days before finally abandoning it. Because they live along a river where there is water, other animals provide social help. Nearby impalas, sensitive to the sight and sounds and presence of predators, keep watch and run when danger approaches. This allows the baboons to scramble into the trees for safety, as they do in the evening to escape potential danger from predators like leopards.

Finding food, seeking shelter, sensitive to the signs of danger and nearby predators, finding solace among the social network, finding one's place in a family or tribe's social interactions, grieving for the loss of family members—all of these traits are also typical of the human primate. The fact that primates have similar social traits indicates that there is a common ancestral source for typical behaviors within the last few million years, much longer than we ever assumed possible.

The theory of natural selection, hinted at by various philosophers and scientists over the centuries like Charles Lyell, Jean-Baptiste Lamarck, Alfred Wallace and Charles Darwin, has now been exhaustively researched and found to be the only explanatory theory for the development of all organic life. Scientists have traced life to 3.85 billion years ago, or about a billion years after the formation of the planet itself about 4.55 billion years ago. Fossil records—and the discovery of eukaryotes, amoebae, microbes, bacteria and DNA—have substantiated the evolution of all plants and animals, including us, ever since.

Scientists estimate that humans and primates had a common ancestor about five million years ago, and that over time differing species of primates, like chimpanzees and bonobos with differing social patterns, evolved in a diffusion of environments. Hominids, relatives of early forms of humans, began using stone tools like axes about 1.5 million years ago.[29]

The oldest known ape-like hominid fossil is *Ardipithecus ramidus* and dates from 4.4 million years ago. Spears appeared about 400,000 ago, and humans, or *Homo sapiens* as we might recognize them and not just as primates, began about 200,000 years ago. The Cro-Magnon paintings found in the Chauvet cave in southern France, visual representations of lions, rhinoceros, and mammoths, have been reliably dated using carbon 14 methods to 32,000 years ago. These artistic wonders are the world's oldest known human art and expression of culture. Our history is much older and richer than we imagined.

Historically, at a time when the earth was young and humans unavailable, the only reality was rocks and gases. Of course, farther back in time there were not even rocks and gases but only a kind of nothingness, the chaos described by the Greeks. Even the chaos noted in *Genesis* is a watery, bubbling kind of nothingness.

Once organisms developed, in that unpredictable random, and chaotic manner it has come to represent for unplanned evolution, the human primate appeared and matured. Now because of massive ignorance, petty self-serving interests, like those of the coal industry, fully adult humans are in a unique position to turn the earth back into rocks and gases whence it began by eliminating or possibly annihilating much of organic life through carbon pollution. Only a few insects, and perhaps cockroaches, will survive the destruction of organic life as we know it if climate change is not altered.

The past is quite well chronicled but is in no way complete. The present doesn't seem to hold much promise. And the future really is unknown, but everyone thinks they know it. As to the origin of life itself, I have no idea how it began nor does any other scientist, though they talk and write about how life began as if they knew. The universe is a great eternal riddle, only partially accessible to the results of philosophy or science. The search is also on for extra-terrestrial life somewhere in the universe, though I believe that search is as futile as the one for eternal life and the fountain of youth.

This brief survey of life on earth over billions of years, and our entry in a primate form over the past few million years, shows our evolutionary progress, and the long time lengths that created the conditions that exist in a few of our primate varietals today. These conditions and evolutions constitute the existing human population in all its manifold variations. DNA markers, now relatively cheap, can follow the human primate journey from its origin about 200,000 years ago in Africa, and the small band who left Africa about 60,000 years ago to enter Europe and breed with Neanderthals, and reveal something about your particular geographic traces and perhaps who your immediate ancestors were.

In the following chapter I'd like to explore further this biological imperative that makes us who we are, and then in succeeding chapters how we engage our primate, genetic inheritances into socialized behaviors.

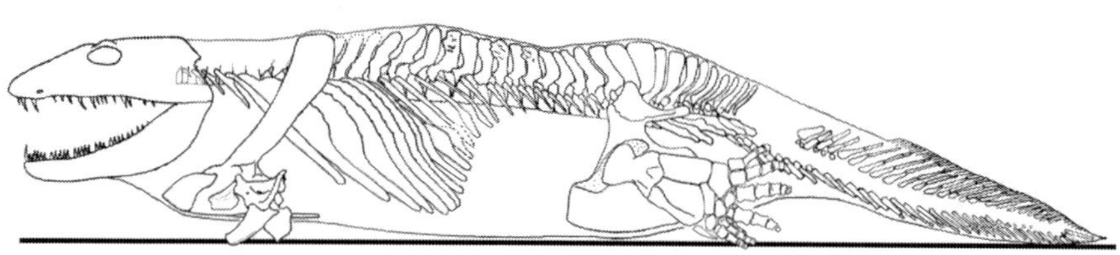

New *Ichthyostega* reconstruction from Ahlberg et al. 2005

Ichthyostega, a tetrapod from c. 365 million years ago.

Pluto, about the size of the moon, at 3 billion miles from earth.

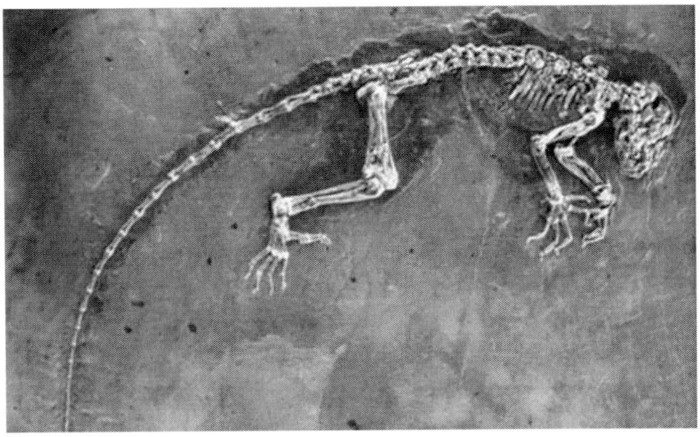

Ida the Ageless, at 47 million years.

Chapter 2

THE BIOLOGICAL IMPERATIVE

We cannot begin to study human primate groupings until we understand the biology of mammals, and the long evolutionary trail that brought us to this status. I'd like to begin with a few, smaller life forms to show the diversity and yet the similarity of life developments that have occurred over incomprehensible time periods.

INSECTS RULE

He rose on two legs and pushed the food source towards his domicile. He had to glean and extract the nutrients from the massive food source, more than twice his size. But he would be able to live off the edible parts for some time.

Such is the life of the various species of dung beetles, admired by the ancient Egyptians who made them a symbol of life. Shit, dung, scat to biologists, or feces more euphemistically is food for many insects and small animals. A dung beetle can transform a mini-ecosystem by disposing of excessive deposits larger animals leave behind because of organic food processing. That waste of larger animals becomes food for small animals is one of the ironies of the ecosystem. In Australia, large quantities of dung beetles have been introduced to help clean up the waste of cattle and sheep and improve the soil quality.[1]

Insects rule. Even small ant colonies have scouts who search for food sources. Sports teams do not have a primate exclusive on scouting as preparation for upcoming activities.

Biologist E. O. Wilson writes about his life's work with insects.

> *Social insects, flourishing at the peak of their 350 million-year history, are the most spectacular in all the great faunas on the land. Their species variously build cities, raise gardens, and conduct endless wars to acquire territory or capture slaves. Their civilizations are built entirely with instincts that guide them through their epic lives.*[2]

We speak about circadian rhythms, a pattern of being awake and sleep modes named after cicadas. Cicadas have a monotonous, high-pitched singing voice intended to lure mates that runs on a 24-hour cycle. We have named them after our disrupted sleep pattern, usually from so-called jet lag that crosses several time zones, as the breakup of our circadian rhythm. But oddly we name it after an insect, not one of our more proximate primate cousins.

Honeybees, as indispensable insects fertilizing flowers and vegetables for human consumption, have been on earth for about 40 million years. Flowers use bees as a substitute for sex, transferring the pollen grains they pick up when visiting a male flower and depositing them in a female pistil flower. But the honeybees are dying out slowly and for easily ascertainable reasons. The reasons may be enough to give our own human species pause. Without honeybees which insect is supposed to pollinate our fruit trees? Will we witness the depletion of all fruit because of the loss of our chief pollinator?

As in most scientific conclusions, the upcoming honeybee extinction has multiple causes. But the principle cause of honeybee reduction includes the widespread use of pesticides. The corollary for the demise of humans is the abundance in the atmosphere of too many carbons. Studying our possible origins is an intellectual exercise. But the demise of our varietal might be less of a mental exercise than one of ultimate survival. After all, like the melting of glaciers, most pre-existing species have simply disappeared into the sea too, sometimes in catastrophic events like that of the dinosaurs, a few leaving their fossilized remains embedded in soft mud that hardened into a stratigraphy of rock that over millennia rose into mountains.

Butterflies too are insects, yet they can see, smell, taste and touch. Their senses however are radically different from human primate senses. Their eyes have scores of what biologists call *ommatidia*, each little eye a receptor of a whole image from a slightly different angle, but all forming a compound image. This is hypersensitivity to the environment.

Charles Darwin saw the impact of how plants and flowers coevolved into a mutual co-existence with insects, relying on each other for the transmittal of pollens and seeds to be spread for reproduction. His book on his study of this coevolution between orchids and insects, a book that sold more copies than his scientific books, demonstrated this symbiotic relationship.

Eyes on the Prize

It is sight, how the eyes have developed, that is an example frequently used by creationists as the principal rationale for Intelligent Design. Scientific research has explained how evolution has perfected sight, that sensitive organ responsive to light. Following the argument from the first chapter on how long evolution takes to make any modifications to its organisms, let's look at some of the existing scientific evidence.

Arendt and Wittbrodt, developmental biologists at the European Molecular Biology Laboratory in Heidelberg, Germany, found vertebrate-like photoreceptor cells in the brains of ragworms, a marine species unchanged for 500 million years. Arendt and his colleagues then looked for genes and proteins that might point to the origin and function of the cells, and hence the origin of eyesight. The ragworm's brain photoreceptor contains a pigment almost identical to the human eye. This is the first instance these proteins, which are key to building the nascent retina in vertebrates, that have been discovered in invertebrates.

Scientists study the lancelet, a thin, sand-burrowing chordate that lives along the bottom of tidal, mostly tropical seas. It is a marine form for tracing how vertebrates evolved. One of those bodily functions is the lancelet eye, actually more like a tube that has light sensors. Vertebrate eyes also have light sensors that travel backwards to an optical nerve where the image is recorded. The lancelet eyespot and the embryonic vertebrate eye in development have similar characteristics leading scientists to believe they both have a common sight origin.

These findings suggest that even the earliest animals had both vertebrate and invertebrate visual systems, and that some of the photoreceptor cells in the invertebrate brain were transformed through a series of steps into vertebrate eyes. Arendt and colleagues concluded that both optical systems existed in an extinct common ancestor called *Urbilateria* and likely arose in its predecessor. Their results suggest the antiquity of both invertebrate and vertebrate photoreceptors. It appears, the researchers suggest, that the two types of receptors was already set in our common ancestor.[3]

The common aspect of all life, insects, rodents, small and large animals, mammals and primates, is that they take in food at an orifice at the front end of their organisms, slowly process the proteins and nutrients they need, then eject waste products the body doesn't need at an orifice at the rear . . . except for the jellyfish, that primordial marine life form, whose mouth is also its anus. This common feature of all life forms, the primal imperative for survival, by itself appears to point to a common organic ancestry.

Imagine for a moment there are no stores, no location where you can purchase food or any kind of eatable supplies. Commercialization only began with the development of urban life about 10,000 years ago. Prior to that, unless you had formed collaborative bands with neighbors, you were literally on your own to find your nutrition, mate(s) and shelter. You had to know, or have been taught, the skills necessary to kill your food, or find where it grew, and then how to protect from theft. Clearly, the strong and aggressive thought it easier to take from others what they were unable or unwilling to find on their own. Women instinctively know this and usually choose the most intelligent or powerful among males who will protect them and their brood. It's as involuntary as a knee jerk.

These and similar examples form the evidence for my argument that we have traits and attributes that were formed millennia ago, and that the development of all organic life and mammalian life led to the evolution of primates and eventually us.

Extinctions

The landmasses and continents that exist today would have been unrecognizable to anyone able to witness them in former epochs. Fossils have been found in mountains that once were shallow seas, or they died in areas that once were riverbeds, or succumbed under mounds of lava ash from volcanic eruptions, or died in the asteroid that crashed into earth 65 million years ago decimating the dinosaurs and other land and marine mammals. One day, unless we maintain its vitality, earth could look like the barren landscape of one of the planets.

Humans cause the extinction of most mammals and species today through hunting and the destruction of habitats. But the most obvious argument in favor of the prevalence of organic evolution and natural selection is not just in organisms but in the evolution of the earth that they inhabited and that supported them. We do not know if species became extinct because of unrecorded astral encounters or because certain species were unable to adapt to new conditions.

Rachael Carson's *Silent Spring* (1962) awakened us to the dangers of ecological pollution and of the wide use of pesticides. Her knowledge of marine life, as chief of publications of the U.S. Fish and Wildlife Service of the U.S. Department of the Interior, gave her public voice in her books and testimony to Congress, prompted a cynical reaction from the chemical industry and acclaim from the general public. Within a decade of her warnings the indiscriminate use of synthetic DDT and other pesticides were outlawed. Congress passed *The Endangered Species Act* in 1974 to help prevent the widespread extinction of animal species.

Carson's work exposed another side to evolution: today's insects evolve very quickly to become immune to the pesticides used to eradicate them. Even many flowers and plants have evolved toxins to resist insects that might harm them. Insects always seem to win in this struggle. Moreover, the liberal use of pesticides harms products all along the food chain. Pathogens too, like the antibiotic drugs resistant to the tuberculosis bacteria, are evolving with ever more improved immune systems. Opening mythical Pandora's box was like opening the fastest evolving organisms—pesky insects and their smaller cousins, bacteria and viruses. Parasites will survive all of us.

More recently, Elizabeth Kolbert's *The Sixth Extinction* (2014) alerted us to the possibility of the next possible extinction—us. She chronicles the five known previous extinctions over the earth's long history that witnessed the likes of the demise of the mastodon and dinosaurs, and then provides illuminating evidence of how species are departing at an alarming rate, their extinctions caused by what humans are doing to the planet, and not because of the collision with another cosmic rock.

Only the last one-sixth of the history of the earth has contained multi-cellular animal life. Although we assume that life moves upward and onward in a predictable, linear fashion, this is not the case. Dinosaurs ruled the mammalian word for more than a hundred million years until their extinction. Had dinosaurs not become extinct they would certainly be on the earth now, and not just in zoos. No other mammal, found everywhere on the earth, so adapted to its environment, yet none survive today. Discoveries reported in the journal *Science* in 2014 indicate that some dinosaurs, known as ornithiscian dinosaurs, had feathers.

One dinosaur discovered in Argentina, with a typical small head, long neck and long tail, was 65 feet long, 30 feet tall, and weighed more than 65 tons. You would not have to have gone to Africa to see one because they lived everywhere on the earth. Then again you might not have been around long to see one, since many happened to be carnivorous. Then again, if you had happened to see one, it would be like an organic jumbo jet with legs—or eight times the size of an elephant—lumbering toward you as it spots you as its next tasty snack.

A dinosaur discovered in Morocco in 2008 was even stranger. Scientists named it *spinosaurus* because it was a swimmer. It was the largest known predatory dinosaur, 50 feet in length, with razor-sharp claws, a conical snout and teeth, and with paddled feet for use in the water. Until the discovery of this unique dinosaur, they were all thought to be land-based behemoths. (See endnote No. 17 for dinosaur citations).

The conclusion to be drawn from dinosaurs, those enormous, mostly carnivorous beasts that surpass imagination, roaming the earth a hundred million years ago, is that if not for the asteroid crashing into the earth that eliminated them, human primates would not exist. Obviously, if humans had existed they would have been snacks for these behemoths. The splayed rows of teeth of a *spinosaurus*, its three-feet long snout, with each tooth about 6-7 inches long, would have been ample enough as a weapon to extinguish humans it spied and pursued with its long strides. The chaotic and completely random experience of a stray lump of extraterrestrial rock careening into earth's trajectory made human evolution possible in the succeeding millions of years after dinosaur extinction.

Similar studies have demonstrated that birds are descendants of dinosaurs. Noah Strycker's 2014 book, *The Thing With Feathers, The Surprising Lives of Birds and What They Reveal About Being Human*, chronicles this comparison.

And while dinosaurs roamed the earth, no other mammal got a larger brain, greater size, or challenged them for supremacy in any way. We owe our existence to their extinction, since there is no reason why they, many of them carnivores, could not still be dominating the earth as the largest mammals ever to emerge from multi-cellular life.

On a more personal level, personal extinction occurs when we die. Everyone over a certain adult age can imagine the possibility of death since its images and evidence exist all around us. Death should not surprise us, regardless of how much we love life. And although we have differing views of what happens thereafter, there is no question that death is permanent and irrevocable. I believe this is how we should view the possible extinction of all of us. Our polluting of the planet and the consequences of the eradication of so many species is enough evidence for the informed, like Elizabeth Kolbert's analysis in *The Sixth Extinction*, to demonstrate that there is an incremental but inevitable death occurring, a slow extinction of the earth's creatures that will eventually include us.

> *Tis our fast intent*
> *To shake all cares and business from our age*
> *Conferring them on younger strengths, while we*
> *Unburden'd crawl toward death.*
>
> (Shakespeare, *King Lear*)

PREDATION AND PROGRESS

A variety of animals have ways of defending themselves from predation. Armadillos and turtles have hard shells. Trees have tough barks and many of their fruit droppings, like the nuts from the buckeye tree—so named because of a large walnut-sized dark brown exterior with a tan center looking like an eye—have tough exteriors and are poisonous to cattle and humans because of high amounts of tannic acid.

Avocadoes have a tough green skin. Eggs have shells. Why is nature so similar in protecting the inner core of its vegetable and animal kingdoms? Can it be that they all have evolved from similar outer protection devices?

Humans do not have hard exteriors. Skin is easily penetrated. So human primates have to find other means of defending themselves against a variety of hurts. They lie, cheat, steal and use verbal behaviors to protect against becoming victims from aggression or intimidation. We reckon these as moral behaviors. But I believe they have their roots in genetic imperatives as a way of protecting the fragile Ego from damage, and from threats that might be more harmful to the Self. Through wits and wariness, humans protect themselves from harm.

By logical extension, states too protect themselves from aggression by other states, or by damaging the image of the state and the social compact among its people, by going to war. This or similar aggravations evolve easier if the dispute is between peoples of differing ethnicity or religion since there is a built-in animosity against "others." Disputes, often historically ingrained, are frequently over territory, or natural resources, or some real or perceived snub.

States act similarly to bands, tribes, clans, packs or prides in the animal kingdom by coming together to protect self-preservation interests by defending rights, image and territories. Karen Hollis, Professor of Psychology, Neuroscience and Behavior at Mount Holyoke College, has studied co-evolutionary adaptations between parasites and hosts, predators and prey in the insect world. Her paper, *Behavior and Learning are Key Elements in the Evolutionary Arms Race*, in an American Psychological Association's Science Brief in May 2015, describes the results of her research with the battles between insects for territory and shelter. All organisms, not just primates or mammals, exhibit similar kinds of aggressive behaviors when threatened.

Like some animals and vegetables, the human inner core protects the organism's delicate and not so delicate seedpods that duplicate its life. These can be very large in relation to the organism itself, and just contain one seed like an avocado or mango pit. They can be visible but abundant like the hundreds of crescent-shaped yellow seeds scattered by each mesquite tree in the summer months in the desert where I live, or similar to watermelon or apple seeds. Or they can be infinitely small, like individual human male sperm of most mammals and primates, and number in the millions with each ejaculation.

Some animals can also be deadly like the Portuguese man o' war first introduced to the world by Thomas Huxley when he discovered them in the tropics as an assistant surgeon while on a voyage to Papua New Guinea. Portuguese men o' war are not single jellyfish, though they look like one. They are actually grouped organic colonies covered with a liquid membrane that resembles an old warship. Their principal organism is really a series of tentacles extending more than 30 feet below the water's surface and filled with venom that kill fish and paralyze the other marine animals that become their food.

One of the first and most extensive organisms is the trilobite, a marine creature that first appeared in the early Cambrian period about 520 million years ago and became extinct about 250 million years ago. They had spines, complex eyes, muscles, internal organs, and appendages: the beginnings of all later life forms. Their millions-of-years life span allowed them to flourish everywhere in the early oceans and become a rich source of a fossilized record of early life on earth. Over time they acquired new forms of eyes and a thinning of the thorax. I was thrilled when I first held a complete trilobite fossil skeleton in Australia in 1984. I recall thinking then that I was possibly holding in my hand the origins of all life on earth. Rocks rule too. Archaeology is the earth's institutional memory recording what happened prior to the arrival of the human primate.

Natural Selection Keeps on Evolving

The earth, like most humans, doesn't willingly yield its secrets. Yet nothing is so intriguing and exasperating as the quest for human origins. The voluminous evidence for the relationship of humans to other primates comes from different sources: anatomical comparisons, and studies of fossils and molecules. During Darwin's day no fossil record supported his claim that primates were related to humans. But now there is an accumulated body of evidence pointing to the 6 million year-old origin of hominids and primates.

According to Plutarch, the Greek historian and biographer, Anaximander (c. 611–547 BCE), a student of Thales and exponent of the evolution of physical beings, wrote: "that originally man was born from animals of a different species." Thus, the indestructibility of matter, the physical adaptation to the environment and evolution (Anaximander said man developed from fish) were ideas promoted in the 5th century BCE, eons prior to Darwin and his theory of natural selection.

Darwin's central thesis is that human biology has evolved by means of natural selection according to random variations in individual organisms and populations. Adaptation is the key to organic change. Natural selection works to make organisms adapt. Genes work within an organism, but natural selection works at the level of individual organisms. The progressive movement over long periods of time starts with genes and moves through organisms to species and varietals. The level of the individual organism fighting to obtain the benefits of self-interest is the same premise that prompted Adam Smith to compose *The Wealth of Nations*, first published in 1776 nearly a century prior to Darwin's *Origin of the Species* in 1859.

Darwin's convincing argument, bolstered by his extensive reading and collection of scientific facts from the vegetative, marine (especially his research on barnacles) and animal worlds (like the adaptive beaks of finches on the Galapagos Islands), allowed him to conclude that each species was not created separately. He concluded that the chief agent of change in humans from other animals was evolution by natural selection, aided by inherited characteristics and the direct action of surrounding conditions.

Darwin argued that there is only a difference of degree between animals and humans since all primates have the same senses for obtaining and processing stimuli. Though certain animals may have limited reasoning ability, they do not have the higher mental capabilities of abstraction, self-consciousness or reflection. Because we have given animals different names seems to presume that they have substantively different natures than humans. But it's difficult to understand, apart from language, how anyone who has kept a dog or observed an elephant can doubt an animal's power of emoting and reasoning. I believe the greatest distinction between us and other primates is our ability to see irony and to laugh, and our ability to tell stories.

The code to understanding natural selection is variability, or the uniqueness of individuals differing in eye color, height, weight, disposition, and all those features by which we can recognize who someone is and differentiate them from everyone else. It is because of the reality of individual variation that we can conclude that subsequent branching into different genotypes is possible, each with slight variations over preceding and succeeding generations. Molecular genetics has unlocked some of these obscured operations of nature.

The rate of change within an organism and between organisms evolving into a new species or varietal occurs much more slowly when big changes in important new proteins are introduced. Moreover, prestigious publications like *The Journal of Infectious Diseases* regularly report of how over 70% of bacterium are mutating into toxic new strains by adapting to live with the antibiotics administered to kill them. Penicillin, for example, no longer stops the spread of gonorrhea.[4]

Viewed from the perspective of microbes, like the lung germ *Mycobacterium tuberculosis*, such microscopic evolution is a daily occurrence. Germs are mutating into a worldwide health problem by discovering new adaptations to drugs designed to counter their rapid and potentially fatal spread.

After reviewing all the evidence, it's not that difficult to grasp that, as far as the universe is concerned, that the presence on earth of humans is the result of a series of developmental biological patterns. Humans, like all other animals, have individual differences or slight variations. These modifications have been acquired through sexual selection and genetic inheritance. But our genetic and cellular proximity to all life forms, but particularly mammals, chimpanzees and other primates, is unmistakable. According to evidence I have read and videotapes I have seen over the years, chimpanzees have killed members of their own species without known motivation. Like human primates, their aggression and occasional violence is often unexplainable.

But because there is an abundance of archaeological and research evidence about the proximity of organisms and primates does not mean the record is complete and satisfactory. The growth of the fossil evidence only supports provisional conclusions about precise lineage. There is a natural tentativeness on the part of scholars in the taxonomy of classifying primates, since finding links between populations presumes there are intermediate forms, the so-called missing links. Thus, archaeologists gave various Latin names to early humanoid fossils like *Homo habilis*, the tool maker, *Homo erectus*, the upright walker, or *Homo sapiens*, the smart one. There are in fact no missing links, only intermediate forms like lineages leading to new species, and possibly the extinction of old ones like the Neanderthals.

From 1.5% to 2.1% of Neanderthal legacy lives on the genetic makeup of modern human primates, indicating that some Neanderthals and Cro-Magnons interbred. Neanderthals dominated Eurasia for hundreds of thousands of years, from roughly 250,000 to 45,000 years ago according to Kate Wong, a senior editor of *Scientific American* who has studied them. When *Homo sapiens* arrived into Europe from 45,000 to 30,000 years ago they overlapped with Neanderthal for over 5,400 years throughout northern Europe, long enough to share, or contest, living spaces. Neanderthals too had symbols carved on cave walls and used stone tools.

A jawbone was discovered in a cave in Romania, radiocarbon dated to 34,000 to 36,000 years ago, was at the time the oldest directly dated human primate, not humanoid, fossil. The jawbone with five enormous molars intact suggests a Neanderthal characteristic and the possibility of interbreeding among the ancestors of modern humans and the extinct Neanderthal strain. Neanderthals have died out as a primate species but still hold fascination as a close primate cousin.[5] It took two years for Swedish evolutionary anthropologist Svante Paabo and his team at the Max Planck Institute to decode the Neanderthal genome. But it was decoded it was clear that Cro-Magnon and Neanderthal peoples interbred and the result is that all human have genes from both group of ancestors.

One current endangered primate, found only on the island of Madagascar, is the mouse lemur. Size is not the primary identification of primates, though humans are the largest of the genus. The huge-eyed, nocturnal mouse lemur could fit in the palm of your hand, but, like all primates, has five digits on each hand. Perhaps the strangest looking primate is the proboscis monkey of Borneo, distinguished by its very large nose and three-foot height. It has webbed feet that allow it to walk upright in the shallow, lagoon waters of the tropics.

The Cat Ba lemur is a raccoon-sized primate found mainly near Vietnam's northern coast and is on the critical endangered list. Today there are only about 55 known to be extant, though the Vietnamese government has created a protected area within a national park to provide them some safety.

The origin of our current species, *Homo Sapiens*, is estimated at 200,000-150,000 years ago, and the dates for mapping the modernity of humans is getting pushed back further in time. For example, 75,000 year-old shell beads strung together for cosmetic adornment have been found in South Africa.[6]

Molecular evidence from chimpanzees reveals that they are more closely related to humans than to gorillas, making a compelling case that humans, hominids, and primates are all related and had a common ancestor. It is the physical brain and human behavior that separate humans from all other creatures.

Chimpanzees (*pan troglodytes*) are our closest mammalian cousins. We shared a common ancestor about 5-7 million years ago.[7] Our genomes differ only about 3.6% according to the latest research, and is closer, at only about one percent, when amino acids are compared. (There are three billion DNA molecules, or nucleotides, in each genome.). As a result, chimps are good models for testing human diseases and conditions. Chimps, baboons, bonobos, orangutans, and gorillas, the other main primates, live basically the same routine. They hunt for food along the ground by day and seek shelter in trees or along the sides of cliffs at night to avoid predators.

Percentage of genes shared with higher primates		
Chimpanzee	human	98.4%
Gorilla	human	97.7%
Orangutan	human	96.4%
Gibbon	human	95%
Monkeys	human	93%

Source: Diamond, 1992

Wildlife sanctuaries have arisen all over equatorial Africa because chimps are in danger from farmers who kill them for raiding their fields, from poachers looking for meat, or from humans looking to sell them as exotic pets. The rapid destruction of their habitats, cleared for timber, mining or farming, also makes them vulnerable to extinction. It would be more convenient to provide necessary services to humans that would mitigate the danger to all African wildlife, to keep soils fertile, seek higher food production yields, and to teach poachers useful skills that would keep them from living off endangered species.

The term "ape" is a mislabeling as it lumps together higher primates without distinguishing major features. Molecular genetics has now revealed more precise differences. Chimps and humans diverted from the line of primate evolution about two million years ago. Put another way, it takes about one million years for one percent of DNA difference to manifest itself in higher primates, usually leading to a new varietal.

But chimps and primates are not the only mammals near us in the evolutionary chain that exhibit quasi-human abilities. The Navy Marine Mammal Program, operating out of Point Loma, California, was created in the early 1960s during the Cold War to detect underwater mines and to assist in locating test torpedoes. The Navy used marine animals, initially dolphins, because they had sonar abilities beyond that of humans. It was known as *Cold Ops*.

In 1977 a team procured beluga whales and deposited them in the Arctic Ocean to detect Soviet submarines. One beluga whale named Noc worked for the Navy for 23 years on a variety of missions.[8]

The amazing thing, according to reports from the Navy team, was that Noc often tried to communicate directly to team members. Noc "spoke" often, both underwater and in the open air, and only around humans. His recordings, clustered around the frequency of high C, revealed patterns that mimicked human speech, but were several octaves below the whale's usual sounds. Noc was trying to communicate to humans not other whales. The conclusion of the team's research indicated that Noc's brain had structures that included self-recognition, memory, advanced socializations skills and language.

There are no single fossils that will demonstrate the evolutionary transition of one species to another. Rather, it is the convergence of multiple, independent inductions from which conclusions become persuasive. For example, researchers like Bruce Lahn, an evolutionary geneticist studying the evolutionary changes in the human brain, have suggested that the brain itself is undergoing major changes. Lahn and his associates found that the proteins encoded in brain genes in primates have evolved at a 30% higher rate than that of non-primates.[9]

The human body is two elementary chemical processes: the chemical storing of substances, and their decomposition. The re-building process continues until specialized cells die like the heart, lungs, liver, and brain. Germ cells on the other hand appear to be immortal as they theoretically last forever. They die only when they are attacked, starve or incur disease. Humans are the long product of billions and billions of generations of cell production. The DNA molecules are the same for all living organisms, even bacteria; only their encoding arrangements differ as they reproduce exact copies of themselves.

DNA is not exactly the same as one's destiny or karma. The environment, like the food we eat, is important too in human primate development. Something called epigenetic information is also housed in our chromosomes and they regulate cellular functions. Certain harmful chemicals—agricultural pesticides top the list—can cause diseases and affect reproductive behavior. These in turn can be passed on to other generations.

While the experience of reproduction may be a pleasurable sporting event, without realizing it, parents can pass on disadvantageous influences to subsequent generations. The gift of life itself, and its abnormalities, could be a burden without the gift-givers knowing what it is they have passed on. But in revolutionary genetic research, it is now possible to have children without intercourse and to exclude harmful genes that lead to lethal diseases.

Primates Preceding Us

The ancestral past of humans has been literally unearthed as archaeologists have discovered more and more ancient humanoid remains. Lucy was discovered in 1974 by my colleague Donald Johanson in Ethiopia and is perhaps the most famous, best preserved and complete, ancient hominid skeletal remains ever found. When she was alive about 3.18 million years ago, she would have stood about 3 feet and weighed about 65 pounds. She is the precursor of modern humans since she shares some of the same biological characteristics.[10]

The discovery of a pygmy skeleton on a remote eastern island in the Indonesian archipelago stunned the scientific and archaeological community when it was announced in

2004. The human bones disrupted time-honored theories about linear progression in human development. The bones belonged to an unidentified form of miniature humans, real Lilliputians, where adults were only three feet tall, who lived on an island called Flores only 13,000 years ago along with pygmy elephants. These people, who evolved in isolation and on an island where there were apparently no predators, were contemporary with *Homo Sapiens*, Cro-Magnon and Neanderthals in Europe. They needed few calories to live. But they had brains smaller than a chimpanzee. These dwarfs are descendants of *Homo Erectus* who wandered out of Africa about 1.8 million years ago to populate the planet. They may have become extinct because of volcanic eruptions about 12,000 years ago although legend has it that they still lived in caves when the Dutch came to this region in the 1500s.

Darwin wrote in *The Descent of Man* in 1871: "It has often and confidently been asserted, that man's origin can never be known: but ignorance more frequently begets confidence than does knowledge: it is those who know little, and that those who know much, who so positively assert that this or that problem will never be solved by science."[11]

Though Darwin was speaking about knowledge about the past based on reflections on reality and scientific evidence, he could have been prescient about the population's knowledge about any subject, including current affairs. A Pew research poll in 2014 showed that more than 70% of Americans scored less than half on a randomized national quiz of topical current event facts that citizens need to be informed about national and international issues. And yet everyone appears to have a mobile phone and computer device that apparently allows each to have access to information only dreamed about in previous generations.

We are all at one with all living life forms. The biological time frame has been plotted with remarkable accuracy by the fossil evidence over the last 500 million years. A tree climbing mammal, about the size of a miniature chimpanzee known as *Eomaia scansoria*, found fossilized in rock sediment, has been dated as 125 million years ago, the oldest known mammal.[12]

No claim in paleo-anthropology can be made without starting an argument, even among paleo-anthropologists. Researching or discussing ancient human ancestors is unsettling, ambiguous and fraught with contemporary bias and imagination, including fantasy movies with special effects. Still, we are always intrigued by new discoveries of these predecessor relatives. The antiquity of humans has surprised even skeptics and the behavioral and genetic similarity with chimpanzees, as noted above, is startling.[13]

Two ancient skulls, one found in the medieval town of Dmanisi in the Republic of Georgia and the other in Chad in central Africa, confirm evolutionary theories of human origin. The discovery shook the family ancestral tree loose of discarded theories and conventional thought about human and humanoid origins.[14] The African skull is seven million years old, or a million years older than any previously recognized hominid, and the Georgian skull is 1.75 million years old, with a brain size about half compared to today's human cranial capacity and enormous canine teeth.[15]

Somewhere in sub-Saharan Africa, either in the great savannahs and mountain forests, about two million years ago, a new species emerged—*australopithecines*, a kind of pre-human primate, about the size of present day chimpanzees. Over a relatively shorter time *Homo Erectus* appeared about 1.9 million years ago. *Homo Habilis*, man the toolmaker—though some primates like chimps also use tools like sticks to dig out ants or termites from nests—appeared about 1.5 million years ago. Each of these predecessors has disappeared and become extinct. Again, it is the vast time frame of these former primates that is difficult to fathom because the experience of a millennia is not within our ken.[16]

So humans, like all ancient primates who have preceded us, are equally vulnerable to extinction, asteroids notwithstanding, as the dinosaurs once were 65 million years ago, creatures that did not have to worry about weight loss. Dinosaur bones have been discovered interred nearly everywhere on earth.[17] Smaller marine creatures like ammonites, appearing about 400 million years ago, became extinct about the same time as dinosaurs.

The origin of *Homo Sapiens* and the fate of Neanderthals have been fundamental questions in human evolutionary studies for over a century. Resolving this issue has been the lack of accurately dated hominid fossils from between 100,000 and 300,000 years ago. But scientists described fossilized hominid skulls from Ethiopia that fill this gap and provide crucial evidence on the location, timing and context of the emergence of *Homo sapiens*.[18] Reliably dated to between 160,000 and 154,000 years ago, these fossils predate classic Neanderthals but lack their special features. They are intermediate between archaic African hominid fossils and later anatomically modern Late Pleistocene humans, and therefore represent the probable immediate ancestors of anatomically modern humans.

Amidst this complexity, no one is really sure if these fossil discoveries are ancestors of all primates like gorillas and chimpanzees, or common ancestors of humans and chimps before they diverted into separate and recognizable species. What is clear is that large portions of the human and hominid record are still missing. Scientists disagree about whether the differences among these fossils can be attributable to individual variation within a species or members of a separate, but related species.

Scientific studies tracing DNA sequences, with slightly altered variations in major population groups, are being traced in cooperative research in human genetics.[19] One particular gene, F13B, a part of a blood-clotting gene and has variations in Europeans, Asians, and Africans. Some genes may help define what it means to be human while others may help in the development of disease-resistant drugs.

Archaeologists like Richard Klein at Stanford believe a neurological improvement in a form of primates occurring about 130,000 years ago in Africa differentiated humans from other close primates like chimpanzees and led to the acceleration of language. The human gene FOXP2 differs from the equivalent gene in chimpanzees and may indicate the breakaway development of language. This special gene was discovered when a large London family with a damaged FOXP2 was found to have severe spoken language difficulties.

Socialization: Know Your Group

Thousands of years ago humans became socialized in bands, tribes or communities and increased their ability to imitate, to reason and to benefit from experience. The socialization was a common ability with other insects, animals and primates. But gradually, over successive modifications, these social capabilities increased. Let's say one man invented a new weapon or snare. Others would learn how to manufacture it and the practice would be replicated in the community. The production of each new art form, sculptured figurine, irrigation method, agricultural technique, use of metal would correspondingly strengthen intellectual capabilities in all who produced or used them. Each new invention benefited trade, success in war, food production, or hunting and mating. People were thus able to propagate children better equipped for survival, and also endow them with inherited superior intellectual abilities and habits which they in turn passed on to subsequent generations. The replication of this process over 100,000 years or more, and definitely within the last 50,000 years, resulted in *Homo Sapiens*.

One of the more contentious and yet intriguing aspects of the evolution of socialization besides advances in technology is altruism. Helping someone for any purpose at all may seem to be the antithesis of self-preservation. Yet I believe that one facet of self-preservation is preserving the group, the clan, the tribe, the extended family. By aiding another an individual is helping preserve the species. I have witnessed in Africa a rhinoceros assisting a nearly dead gazelle mauled by a lioness by dragging the animal away from danger, so there must be even some cross-species altruism among animals. Any form of cooperation can lead one to conclude that group survival is also an evolutionary trait.

Until new archaeological discoveries occur, our historical reconstruction of the cultural development of our human primate species will be speculative. There had to be a minimum of cerebral development to learn to outwit other primates in food sources and sustainability. Humans learned to process and eat wheat as a primary food about 10,000 years ago in places like Jericho and Aleppo, two of the oldest urban settlements in the world. As in all cultural pursuits, the agricultural revolution, like the scientific, did not occur in a major blast but more like a series of creeping modifications.

The absolute necessity of living by means of vegetation is solidified in the most ancient myths. These demigods and goddesses began in ancient Sumer. But we are more familiar with the Greek legends, beginning with Demeter, sister of Zeus and mother of Persephone. Demeter was the goddess of grains, grasses and fruits. Her hierarchy in the Greek mythology demonstrates her importance to the agricultural life of Greeks. Dionysius or Bacchus was the god of the grape and wines, a fruit that emerged as a favorite beverage, one that has not lost any ingestible relevance.

The hunting skills were easier to mature and were essential to survival of the immediate family and the tribe or clan. I recall vividly the scenes from the four cave paintings I visited in France years ago. The extraordinary Paleolithic illustrations, at least 13,000 years old, of the cave at Niaux in southern France next to the Pyrenees were perhaps the most awesome. The tour guide led me into the interior where I experienced total darkness and complete isolation. The absence of any light, the fate of blind people, makes anyone instantly fearful.

At an irregularly shaped amphitheater probably about a half-mile in the interior, our flashlights illuminated the panorama. Horses, reindeer, lots of bison (actually now extinct aurochs), ibex and a few unidentifiable, probably extinct animals delicately graced the cave walls. Often the unknown artist incorporated the curvature of the walls into the bellies of the animals. And they were depicted by these first human primate artists were magnificent nuances of symmetry and perspective and in various modes of action and movement.

Scholars assume that these cave paintings served as ritualistic and technical introductions to the hunters of the hunt for these animals. But I'm not so sure. I think that possibly the artistic expression itself, a manifestation of cultural development, was itself the motive, and not just the practical exercise of hunting, important as that was to survival. Here in these French caves artistry was born as a significant human and cognitive trait, and not just the more obvious skill of learning how to kill for food.

An analogy with one animal species, the dog, will serve as an illustration of the wide diversity of our genus over time.

The *canis familiaris* is the common genus name for the domesticated dog, ranging from the small, trainable and loveable Maltese to the Doberman Pinscher. Recently, mitochondrial evidence has added to the persuasive evidence about heritability. Over the years since dogs were domesticated, they have evolved to fulfill a variety of human uses, including herding sheep, guarding the property, pulling arctic sleds, and warming laps.

Despite the vast differences in size, color, appearance and disposition, they all have a common origin with wolves, coyotes and jackals, their un-domesticated cousins. Molecular biologist Peter Savolainen of the Royal Institute of Technology in Sweden conducted a study of 654 domestic dogs and concluded that their DNA is traceable to a single gene poll from wolves living 15,000 years ago.[20]

Once domesticated, dogs will follow their pattern of behaving submissively to the Alpha male or dominant female as they would if they were in a band or pack. But if once liberated from that environment, and if grouped together, even playfully as a pack, they go on the hunt as their genes command.

For instance, on July 10, 2014, a 72-year-old woman was attacked by a pack of dogs in a Phoenix park while volunteering to collect discarded cans as she had done many times in the past. A passer-by witnessed the attack and called 911. When police officers arrived, one of the dogs attacked one officer who had to fire his weapon hitting the dog in the mouth. It is not unusual to see packs of dogs, which may not be feral, roaming the streets of major cities. Dogs, like humans and other primates and animals, will revert to their natural aggressive state for survival purposes if conditions warrant, even if they have been domesticated by humans and act placidly and non-aggressively when in their presence.

Dogs (*Canus lupus familiaris*) are only one of the animals the human primate has domesticated. Cats too have become popular pets, but they have been only partially domesticated, and they will never become a human's best friend no matter how much they purr in our laps and appear to be docile. The less familiar understanding is that the human primate has also evolved to its changing environment and has become more accustomed to our smaller, non-primate cousin relatives. The non-domesticated bigger cousins of cats, like panthers, tigers, mountain lions, wolverines, also hunt alone and not in packs, so do not generally exhibit group hunting behavior.

Only humans, of all the approximately 5,500 mammals, have domesticated other species. The most significant has been the cow, as we receive so much in return from this one animal–—meat, milk, all diary products, and leather, among other byproducts. Possibly the next most important is the horse, the earliest transportation vehicle. After the horse came sheep, goats, pigs and chickens, all food sources.

Darwin's assembly of known zoological facts, and his own experiments, led him to the conclusion that humans like all living things were derived from ancestral organisms. "I have now endeavored to show that some of the most distinctive characters of man have in all probability been acquired, either directly, or more commonly indirectly, through natural selection."[21] However humans are zoologically classified, they are the result of continuous biological variation from existing and certainly extinct mammals and progenitor life forms.[22]

Otzi, the Ice Man

On September 19,1991 a group of hikers came upon a rare find. A human corpse had been entombed, "en-iced" might be a more appropriate term, in an Alps glacier where he had lain for nearly 5,300 years. The mummified man had lived in the Late Neolithic period. He was 5 feet, 2 or 3 inches in height, and was between 35-40 years old at death. When he died, probably of exhaustion or hypothermia high on the glacier, he was wearing leather clothing, his shoes were stuffed with hay, and he was carrying a copper axe, flint knife and bow and arrows, the very stuff of Neolithic survival. He was probably a shepherd attached to a Neolithic agricultural community on the Italian side of the Alps during his lifetime.[23] He wore leggings, a loincloth, an upper garment of fur, leather shoes stuffed with grasses, a grass over-cloak plaited with sewn grass, and a conical fur cap with a chin-strap. He had a spear point embedded in his hip and had broken ribs and a fractured skull.

He carried with him everything he needed to survive alone and in the mountains including:

- bow-stave made of yew wood with a quiver and 14 arrows tipped with bone
- copper axe (neither bronze or iron had been invented yet)
- back-pack or pannier made of hazel wood and lashed with cords made from twisted grasses
- birch-bark containing embers for making a fire
- flint dagger lashed with animal sinews to an oak haft and stitched grass scabbard
- calf-leather fanny-pack or belt-pouch containing bits of ember
- blade scrapper for sharpening flint tools
- drilling tool made of flint
- awl made of animal bone and globs of pyrites for starting a fire
- net for catching birds
- loose cords of twisted grass probably as repairs for existing equipment

Ice Man also carried strapped to his wrist clumps of birch fungus probably as his medicine kit. He had an astonishing knowledge, lost in our modern civilization but indispensable in the Neolithic, of a wide variety of tree species, as each type of wood in his equipment has specialized purposes. For example, he used hazel wood for its flexibility and light weight for the backpack; the evergreen, conifer yew for its lack of resin, its toughness and highly elastic use for his bow stave; and ash for his dagger handle. He ate the fruit of the blackthorn tree for part of his diet. He used the resin of the birch made into tar as his glue.

A person of this age would have to possess all the knowledge and skills necessary to survive personally and as a member of an agricultural and hunting community. This would include knowledge of hunting, fishing and food-gathering, agricultural tasks, knowledge about domesticated animals and livestock (the dog, goat, sheep, pig and cow were domesticated, but not yet at Ice Man's time the chicken or horse), how to build huts and fires, and what trees, shrubs and grasses would be useful in all situations, and how to find and work flint. Ice Man's education would have consisted of learning all these survival techniques, and his inclusion into the community may well have depended on demonstrating such knowledge.

Ice Man is the oldest complete corpse existing. Egyptian corpses are mummified and missing internal organs. What is remarkable is that, unlike other examples of human remains of that antiquity, he has been preserved naturally in ice, high up in the Alps un-disturbed for 5,300 years. He is not the oldest human relic, but the fact that he is completely intact makes him a valuable ancestor for biological research. He lived in Europe about the same time as the Sumerian civilization at its height, before the rise of the Egyptians, and prior to the establishment of the Kingdom in Israel, biological proof of humankind's long journey to the present.

Ice Man is not the end of the story, and neither are we. Evolution is not a linear, rigid structure that can be easily identified and described, but a series of small, incremental changes over very long time periods. Evolution is more like the Internet, an informal, huge web of flexible connections capable of reshaping itself perpetually. Add a million years to that process and, barring a cosmic catastrophe, future archaeologists may not be able to recognize which branch of the human tree, if one still exists, we came from.

Survival instincts passed on through multiple generations are within us today. They include aggression for hunting, the search for food—somewhat suppressed in the developed world where it can be easily purchased—the search and fight for a mate or mates, and the suspicion of others who want to take what we have accumulated. We think of these traits as primitive, and indeed they are. But they are who we are. These instincts came from all primate forms that have gone before us because they are essential for survival.

The literary world is filled with narratives about how people survive in difficult conditions. I think of Ernest Hemingway's 1952 classic work of fiction *The Old Man and the Sea* for individual validation of courage by an elderly man facing death alone on the ocean, and of William Golding's 1954 classic *Lord of the Flies* about a group of boys stuck on an uninhabited island who revert to animal behaviors like a pack of feral dogs. Both authors received the Nobel Prize in Literature for these artistic masterpieces of survival.

KENNEWICK MAN

Kennewick Man, the 9,000 year-old adult male found on the banks of the Columbia River in 1996, is the oldest skeleton of a human primate discovered in North America. Unlike Otzi, the Ice Man, who was frozen and found more or less anatomically intact, Kennewick Man was only a skeleton.

A forensic analysis of this amazing find revealed that he was more Polynesian and not at all Native American in his features, a man more related to early Japanese inhabitants like the Ainu than to the Mongolian.[24] This conclusion alone indicates that the migratory patterns of the early inhabitants of northern North America arrived either prior to, to a little later than, immigrants from separate Asian ethnicities.

Kennewick Man's remains had to remain literally in limbo while a lawsuit dragged on for more than a decade. The Umatilla Indians claimed that he was one of their ancestors and had to be buried according to their burial customs. Legally, they were correct, as they were following the *Native American Graves Protection and Repatriation Act* that prohibited the desecration of Indian gravesites and violated Indian religious beliefs. However, in a contentious court ruling after several years, the presiding judge ruled that scientists should have an opportunity to examine this most precious find. The judge also ruled, based on forensic evidence, that Kennewick Man was much older than the Umatilla tribe and therefore not one of their ancestors.

The concluding evidence is that Kennewick Man does not correspond to "any living human population."[25]

But there are older skeletal remains, and the story of who was the first American keeps getting pushed back further in time. Archaeologists named her Naia, after the water nymphs of Greek mythology, and she had died as a teenager. Her skeleton was found in a cave in the Yucatan in Mexico in 2007 and was dated from 12,000 to 13,000 years old, the oldest human bones found in North America.[26] The facial markings from the skull reveal slight differences but the genetic evidence supports the conclusion that these early American came from Asia toward the end of the last Ice Age.

The search for where we came from is ongoing. Speculation about where we go next, if anywhere, is ongoing, but essentially remains a puzzle except to theologians. Perhaps the author of *Ecclesiastes* had it dead right: "For what befalls the sons of men befalls beasts . . . as one dies, so does the other . . . all go to one place; all are of the dust, and all turn to dust again." [27]

Comparative biological searches, primarily between human and chimps, our closest DNA biological cousins, are also ongoing research project. Although the brain capacities are divergent, and there are exceptions, chimps and humans share some of the same kind of cognitive functions.[28] Over time chimps too have evolved beyond their earlier ancestry. Young children and chimps perform equally well on general IQ tests such as spatial and quantitative measures.[29]

Our evolution continues, but with some misgivings. We have emerged, not just to stake out some territory like our mammalian relatives, but to dominate the whole planet, all its land and uses. Our only rivals are ourselves, other real human primates. But then through ignorance, careless neglect, wanton disregard, and willful negligence, we are presently in the process of eliminating many of the species, or other fellow primates, that unconsciously allowed us to evolve into existence. And now through over-population and the unregulated use of nature's resources, like unbridled adolescent equines, we are now destroying the kindness of earth's bounty in order to enrich and entertain ourselves.

The available evidence, and not just from primate archaeology, for the long biological development of the human primate is quite overwhelming. Researchers have analyzed worm, fly and human genomes and have discovered a number of key, common genetic processes.[30] This can only mean that all organic life has a shared ancestry, whether or not we believe we have a lot in common with worms and flies. The similarity of some genes also means that researchers can locate some genes that can help cure or mitigate diseases. We've come a long way, but no one knows how far we have yet to go.

Though we don't know how long we will endure, we do know something of how far we have come cognitively, especially in artistic representation, and in consciousness of external reality and the ability to re-create it. Imagine being able to walk into the newly discovered (in 1994) Cave of Chauvet-Pont-d'Arc in southern France (or at Lascaux, or Altamira in northern Spain) and viewing 442 animals painted on the walls, or see stenciled handprints 40,000-15,000 years old in several caves in France.[31]

Scientific teams in 2014 at Harvard and the University of Copenhagen analyzed the largest data set yet of European DNA in excavated skeletons and discovered that different migrations merged in Europe to consolidate the current population. The first group arrived about 45,000 years ago and these were the Cro-Magnon peoples. Farmers and agriculturists arrived about 8,000 years ago and the two groups lived apparently side by side and with unknown aggression. Then about 4,500 years ago a group of nomadic sheepherders entered from today's western Russia. Not only has Europe experienced a cultural mix but a genetic one as well.[32]

The length of human primate artistry is incredibly long. Here is a brief summary of that archaeological record of human artistic representation:

100,000	Blombos cave, South Africa, symbolic representations, shell beads, body ornaments, etc.
75,000-65,000	Blombos cave, South Africa cross-hatching etchings
40,000	Hohlenstein-Stadel cave, Germany, half-human, half-lion foot-tall figurine
36,000	Cave of Chauvet-Pont d'Arc, horse paintings
35,000	Venus figurine, Hohlen Fels cave, Germany
25,000	Lady of Brassempouy, facial bust in SW France
19,000–15,000	Altamira Cave in northern Spain

Such symbolic and artistic examples demonstrate the long history of our evolving consciousness of who we are, our use of natural artifacts for adornment, our preference for creating expressions of natural objects, and our affinity with the animals we hunted, and the human relationship with them as food and living inhabitants of the land.

> *And the life of man, solitary, poor, nasty, brutish, and short.*
> (Thomas Hobbes, *Leviathan*).[33]

So although our primate ancestry is relatively recent, in the thousands of years, our organic life is even more ancient, in the millions of years. All organic life forms exhibit some similarities, known as homology. Human arms have a humerus, radius, ulna, five carpals, five metacarpals, and fifteen phalanges . . . as do bats and birds.[34] You may not feel much of an affinity for bats, but you need to explain to yourself how it is that your arm structure is so similar to that of one flying creature you fear, and others you exult in. We are the end results thus far of the very long evolutionary process.

Chimps waiting for visitor food, Islamabad, Pakistan.

Venus fly trap: The carnivorous plant.

Chapter 3

THE PROCREATION IMPERATIVE

One of the most overused words in contemporary language is fuck, used foremost as a noun and a verb, and universally in current commercial entertainment as a perpetual swear word. But its original use was none of those. It was an acronym used by London policemen in the 19th century to abbreviate the statute against the arrest of prostitutes: Fornication, Uncommon Knowledge, or FUCK for short, to be entered into the arrest books. Its ubiquitous use now is an example of how keenly humans value the ability to procreate, or for personal recreation and pleasurable entertainment. It is clearly now meant to refer to all sexual activity—that rewarding duet of the dopamine in our brains—and not just the arrest of British prostitutes.

The primary purpose of all life, vegetative and animal, is duplication of itself, and to maximize the survival of the species, not just the individual. We live to fulfill the lives of our eggs and seeds. I call this instinct the biological imperative because all sexual impulses originate within us, as they do in all organisms, an internal compulsion to perpetuate the species. For humans, sex is a temporary insanity and mystery curable in part by marriage and partnerships. But its physiological insistence is a reminder of our ancestral origins.

As I noted in one of my earlier books, *Lords of the Scrolls*, *Literary Traditions in the Bible and Gospels*, biblical scholars admit that there are two creation stories, *Genesis I* and *Genesis II.* In *Genesis I* God creates in six days sequentially, and on the last day creates humans, "male and female he created them." In *Genesis II*, God creates the earth and the heavens, then creates man, then the plants and the animals, and finally woman from man, a totally different order of organic and human creation than the earlier version. Did God then create humans together, as in *Genesis I*, or separately and at different times, as in *Genesis II*?

The merger of two creation myths, or the carelessness of editors, says as much about editing the language of creation as it is confusing about creation itself. The irony is that belief in creationism has its own biblical ambiguities about the priority of human origins.

Each gender is born with its genitals that correspond to the opposite gender's genitals. Women's genitals are hidden, and men's are exposed, as if they were a biological engineering malfunction: not fitting inside anywhere they were positioned on the outside almost as an afterthought. I reiterate this obvious fact only to point out that the result of our gender and its operating features has nothing to do with individual choice. It is determined not by so-called human free will but by the imperatives of cell development. We may believe we have absolute freedom over our behaviors but we have no freedom over how we came into existence, or how our genes rule and compel us.

Sex likely began in ancient plants, some of which have stamens, pistils, and reproductive organs. Some plants pollinate just so they can disburse their seeds or seedpods. Animal evolution chose the path of sexual dimorphism whereby two distinct genders had to copulate to reproduce the species. The human primate is the result of this long, long evolutionary development.

Welwhitschia mirabilis, the native plant of Namibia, is possibly the oldest known plant on earth, dating from the late Paleozoic era, or about 260 million years ago. It is a bizarre-looking plant that can live typically for 500-600 years. An existing one in Namibia is 2,000 years old. But this is the sex part—these plants have both male and female genders. The male has a pistil-like structure, and the female a cone-shaped stigma. Both male and female plants have nectars that attract pollinating insects. The diversification into genders is uncommonly old, and even among plants it takes a turn into maturity for its duplicating tendencies and properties to emerge.

We can start understanding who we are in this chapter with the body in all its physical, tactile and psychological characteristics because it is the most immediate sense of ourselves. Some of the descriptions are clinical but serviceable as tidbits of knowledge about how we function. In the following chapters we'll explore the mind, the psyche and social activities as other avenues to understanding who we are.

But it is important to note again that the body is composed of trillions of single cells. Single cells can be collections and toolbox kits of composite cells that have specific instructions for organizing and replicating each body part, like eyes, livers and lungs, bladders and brains. How this all came together is the amazing marvel of evolution and natural selection, all originating with the first single cell millions of years ago that started the organic evolutionary process. Understanding this biological beginning, and its evolution and continuation through sexual mating, is elementary to exploring all human actions and relationships.

So a person's life is actually the life of cells combined and extended. This miniaturization of life forms, and its ancient history, would be my lifelong fascination. We are still trying to figure out cells that can kill us, like cancer cells. But sex is just as old, and the second most compelling impulse we have for survival after food.

SEX: THE GENDER DIFFERENCE

In the archaeological site at the Messel pit near Frankfurt, Germany, two fossilized turtles were discovered recently *in flagrante delicto*, in the act of copulating. It occurred 47 million years ago.[1] They were probably mating at the surface of a volcanic lake then dived in to avoid disturbances. Unhappily for this couple, volcanic lakes are typically filled with poisonous gases that dissolve easily in water. Dying for sex, usually not during sex, often comes at a lethal price. But the pair did not disengage because of this danger.

So organic sex certainly existed in the Eocene period 34 to 56 million years ago, and probably a few million years prior. The essence of nearly all forms of organic and certainly mammalian life is gender, sexual dimorphism. The male and female have to mate somehow to create new duplicates of the same species. There is no other formula, except for hermaphrodites, an exception to the rule. We don't emerge as hybrids. With only exceptional instances, we are born as either male or female.

Some consider the human species, labeled *Homo sapiens sapiens*, a little less than the angels. But we are really just a little more than the simians and primates, and a form of mammalian life that is hundreds of millions years old. So when we go to the root of human existence we unquestionably encounter animal existence. All other suppositions are cultural and mythical accretions, cosmetic and subtle rituals that lead to the mating game, such as dress, music, dance, or body ornaments. All other survival activities are ancillary to this primary function of all life—to reproduce itself, to replicate DNA, the origin of which is at least two billion years old in the primate species. Other so-called human activities, even occupations, are distracting interludes.

The campfires, the caves, the clubs, the discos, the bars and taverns are a few of the venues where partners meet, socialize, and plan encounters. The results may not lead to fertilization, but the experiences and experimentation are useful, and results usually satisfying. Sex is what we were made for because it is how duplication, the foundation of all organic life, occurs. All human actions are directed toward survival, and genetically desired by all the combined cells comprising the human body. The compulsion to duplicate itself is the single goal of all organisms and all the cells that compose them. All animal and human emotions are rooted in genetic inheritances, and are responses to environmental changes, including the actions of others in the same species.

Without sex, we, like the Neanderthal other extinct primate relatives, would also become extinct. With sex, the unity of the species has dominated the earth. Yet it is on the way to depleting all available resources necessary to sustain its existence. Having over-run the planet in a relatively short time, and relative to the history of mammalian life, sexual productivity is hastening our demise through over-population. But sex is not only for procreation. It is estimated that 30% of all data transmitted over the Internet is porn.

Ancient literary authors invented semi-divine heroes and heroines to show that life could be elevated beyond the mundane hardships of ordinary life. The Greeks portrayed gods and goddesses having intercourse with mortals and thus passing on a semi-divine status on certain individuals. The Greeks had Hercules, Perseus, Theseus and Odysseus as part human and part divine heroes. Biblical narrators did the same for their heroes, as Adam, Moses, David, Joseph and Samson were all endowed with superhuman abilities. But we must differentiate fiction, created to elevate the ordeals of the human primate to an idealized state, from its actual biological origins.

When offspring arrive we coddle them because they are literally extensions of ourselves, reproductions of our organs, a process that dominates all life forms. Discovering the code of built-in replication is so new to us that its mysteries are even more challenging that the first questions about existence, essence, substance and being that tantalized and teased Greek philosophers. The young and immature are little bundles of cute but also of primate instincts: curious, excitable, exploring and acquisitive. After learning "mama" and "papa," the next uttered words are usually "no" and "mine."

The hunt for the building blocks of matter torments physicists, as does the immensity of the universe amaze and bewilder the casual observer, as does the mystery of sex in the normal human. If you have to ask why sex is what it is, you haven't appreciated in all the previous epochs of your DNA, or where we've evolved from.

Anatomically, male sperm and female eggs are germ cells and exist independently outside the embryo. During fetal development they migrate to the gonads. During the earliest embryonic stages they are indistinguishable as either male or female but soon differentiate into testes or ovaries. Scientists are interested in studying how cells give rise to sperm and eggs, how they differentiate into sex, and how mammalian sex chromosomes initially evolved.

Technically, sex is achieved through meiosis in which a parent cell divides producing cells with half of the genes from each parent. The zygote is, however, genetically different from either parent though there are many common traits and facial resemblances.[2] Males have an X and Y chromosome, and females have two X chromosomes. This same determination of sexual differentiation is in all humans, most mammals and some insects and plants.

Knowing the high significance of sexual differentiation in its biological and psychological contributions to human life, isn't it important to fully educate the young to its intricacies and consequences? Heated debates still occur in school districts about sex education programs. Many parents want to eliminate selected sexual-related topics like abortion from textbooks. Arizona, for example, does not require school districts and charter schools to offer sex education. Some school districts offer no sex education at all, thus forfeiting adult responsibility to educate the young in life's most important kind of knowledge.

My parents never instructed me in anything about sex. My grandmother took it upon herself to say a few ambiguous words to me when I was about twelve, but they were far from comprehensive. Regardless of squeamishness or religious hesitancies, parents should instruct the young primates about life's chief method for reproduction, its genetic inheritances, biological imperatives and drives, and consequences for mistakes.

Came something saucily into the world before he was sent for.
(Shakespeare, *King Lear*)

Male infertility has been identified as depletion in the Y chromosome. Technology now exists whereby male/female reproduction is un-necessary to create living organisms. They can be produced in a laboratory from stem cells and clones made from other living cells.

I think that when men see a human female form, that although sexual attractiveness is certainly present, that the wider hips, moderately enlarged buttocks and breasts are an ancestral stimulus that indicates that this woman will be a fertile and a successful mother of children. Fertile women have a waist to hip ratio of between 65 to 80 percent, which decreases slightly as they age.

The affinity men feel for female breasts and their relative size is a visual cue, first of all, of the presence of a female that typically stimulates a sexual response, and secondly that such an individual can feed more than one child. The sight of breasts is as much a sign of maternity as it is just a mere sexual attraction. Even Darwin wondered why men are born with nipples. Paternity, and not just the immediacy of sex, is also compelling as an instinct when men see human female forms. The earth's human population has quadrupled in my lifetime so the schema must be working. Yet there has to be a limit on the delicate balance between the number of humans the earth can safely maintain and the resources that sustain them. We just don't know what that ratio is. As John Stuart Mill noted in 1848:

If the earth must lose that great portion of its pleasantness which it owes to things that the unlimited increase of wealth and population would extirpate from it, for the mere purpose of enabling it to support a larger, but not a better or happier population, I sincerely hope, for the sake of prosperity, that they will be content to be stationary, long before necessity compels them to it.[3]

COUPLING: PARTNERS IN PROCREATION

The key to copulation is the choice of a male by a female who, in the animal world, must make a decision based on of a variety of visual cues—the brightness of the feathers, a larger part of the anatomy, a longer jump, a faster takeoff, or a less likely chance of passing on diseases. A female's choice of a mate is not easy, especially among humans. The males may sing, bellow, or cry out to attract females (sounds like a rock band's antics), but she, queen of her species, must decide among a possible assembly of suitors. And her choice is unerringly based on a variety of inherited cues that differentiates one among many. The next generation, or her very survival, depends on the carefulness of her decision.

In my view, the unity of a couple, however, does not depend on the blessings of a cleric or a religious representative, or a binding contractual relationship with the state. It is a mutual understanding between the man and the woman (or of a same sex couple, or even multiple partners) of the love for each other. My fiancee and I have a permanent bond, not cleared by a third party tied to, and needing to verify, our love. If we are free to choose our partners, we are free to choose the nature of the relationship without governmental or religious concession.

Marriage statutes today are linked too closely to government monetary benefits, such as social security earnings. In the 2013 *United States v. Windsor* case the U.S. United Supreme Court struck down the *Defense of Marriage Act*, a pointedly political act to sabotage gay unions, as unconstitutional, and in 2015 decided that gay unions were constitutional. Many states do recognize common law marriages, which is acknowledgement that the states really have no need for sanctifying personal unions.

Marriage rates have been falling for decades and very likely mirror declining economic opportunities and falling wages. In 1960, according to the Census Bureau, the marriage rate in the United States was 72.2%. By 1980 it had fallen to 62.3%. In 2012 it was 50.5%. There is a strong correlation between marriage and wealth, and compelling evidence that married couples enjoy better income levels and savings than singles. Educated women defer marrying less economically appealing males who earn less, have higher debts and negligible savings. Cohabitation with partners is trending.

There is hope for older adults in the golden years for whom sexual activity does not lead to procreation. According to the latest research "Marital characteristics and the sexual Relationships of U.S. older adults," in *Archives of Sexual Behavior* (2015), married couples in retirement years have an increase in the frequency of sexual activity. At a time when the stereotype of older, long-married Americans is that they have faded in all health and especially sexual activities, the research indicates that sexual activity and intimacy actually increases the longer the couple has been married.

Religious views can also color the coupling. The Mormon Church renounced polygamy in 1890 though many families, and as a regular practice among Mormon splinter churches, continued to practice it but under the radar for more than a hundred years. Nevertheless, the

Mormon Church still teaches that a man can be sealed (the exact definition of which is ambiguous) in the afterlife with more than one wife he had in life, but only if such a seal was conducted in a temple ritual. So officially in this life for Mormons there is no polygamy, but there is hereafter. Mormons marrying non-Mormons do not have this seal deal, because the church does not sanction the ritual for them. Moreover, there is no official church policy on whether this privilege in the hereafter applies to gay couples.

Among human primates the gender choices are first of all visual attractiveness, beauty for the woman, handsomeness for the man, and then all the paraphernalia that usually revolve around the resources—for the most part money, but also educational level and character—one or both brings to the union. In the end, however, the sexual imperative ultimately is about how you nurture the next generation to carry on. The attractiveness of the genders for each opposite gender is palpable and leads to romance, and usually to increasing attachment, bonding and love.

To create the feeling of power, individuals elevate their persona with visual cues. Men tattoo their bodies with signs meant to convey power or significance, or to terrify the enemy. Women adorn their bodies with jewelry and elaborate cosmetics—lipstick, eye shadow, moisturizers, mascara—to attract the opposite sex visually, though for the most part their bodies—breasts, wide hips, enlarged buttocks, color and styling of head hair—hold the major attractiveness to men.

The human primate will spend an inordinate amount of time and energy on learning how to navigate relationships with the opposite gender. This goes far beyond magazines, TV shows, mother/daughter and father/son talks. Learning about the physical differences will be the easiest part of discovering gender differences. The hardest part will be in understanding the motives and inclinations of the opposite gender, finding appropriate responses to verbal or non-verbal cues, and then revising any patterns developed. The basic rule is that there are no set formulas, and plenty of surprises awaiting the uninitiated.

A few humans can seek to forfeit the sexual impulses for other motives. As a youth and a Jesuit seminarian—we priests-to-be thought of ourselves as God's marines—I had a vow of chastity (also poverty and obedience) for ten years, from the age of 18 to 28. (Coincidentally, when Charles Darwin began his journey as a natural scientist on *The Beagle* at age 22 he was a devout Anglican.) I was a monk in the old monastic and ascetic tradition because I was making an offering of my body for a higher good and as a service later to others. What I learned later in life affected me profoundly because, like Darwin, it literally overturned my religious worldview, which, I believed, could not be sustained knowing the archaeological evidence that impacted religion, history and science.

The monastic life of study and contemplation was idealistic but also altruistic, both feelings we all have from time to time, but not to the same limits of personal sacrifice to control one of life's most celebrated impulses. Though we live in the Information Age when computers can retain and speed information anywhere in a heartbeat, there are still those who believe that looking within, contemplating and reflecting on what the brain holds and intuits, is an essential and worthwhile human activity.

Then again, in many ways all humans, primates and social animals, control their passions to conform to the social behavior of the group, tribe, or clan. Young males are especially aggressive, but will control their aggression in the presence of the dominant male (or gang leader) and thus become more acceptable to fit into the group. Like monks, this too is a form of genetic obedience to the dominant male, like God, or the dominance in one's life of a divinity, a father or elderly male, or noteworthy authority.

I believe that a recurring feat experienced during times of intense aggression like war is an evolutionary trait we share with all primates. The experience of returning soldiers often results in Post Traumatic Stress Syndrome (PTSD), an anxiety consequence of fear in battle. War is a time when intense male bonding occurs, but also a time when it is possible to die quickly. When the effects of combat on veterans are extreme, suicide is more likely on returning home.

The primate will to survive is not only individual but collective. Often the group, usually composed of males, have to bond to survive together, and this perfectly describes men in combat. Thus, male bonding and aggression together form two evolutionary-derived traits. When the effects of the loss of bonding in an environment like combat, and the lingering effects of war are relieved, a mental disorder like PTSD results.[4]

Nesting: Nurturing the Young

Charles Darwin noted in *The Descent of Man* in 1871:

> *We have now seen that man is variable in body and mind; and that the variations are induced, either directly or indirectly, by the same general causes, and obey the same general laws, as with the lower animals.*[5]

Needed: a safe place to mate, then a comfortable place to raise and feed the family nutritionally so the children can live, which they then in turn mate and feed on their own as adults. Sounds like humans. But it is also the short life cycle of mosquitos.

Males and female mosquitoes mate and the females feed on vegetative life for energy. They supplement this with protein from actual blood—animal blood usually, but often human blood when they can find it—to enhance egg production. Mosquitoes lay their eggs in standing water and the larvae hatch in about 48 hours. Soon, the larvae mature into pupae and find their own vegetative nutrition. They develop quickly into adult mosquitos and the cycle starts over. We think of them as annoying. But they too are just trying to survive and propagate. Mosquitoes, as insects, cause more human deaths that any other animal predator.

Do we call this *insect nature* as we call acts humans do as a part of *human nature*? Usually we say what insects do is instinctual, what Thomas Hobbes called "appetite or desire." What insects do behaviorally is not that different from what humans do in their activities. All organic behaviors are based on instincts and embedded genetic propensities and how they behave in certain environments.

Mayflies—sometimes known as *ephemeros*, after the Greek word for literally lasting one day—have only lives of one day, or sometimes for just a few minutes. In the summer month they arise from the water, mate while in mid-air, drop their eggs back in the water, and die. The cycle is then repeated the following year. A short feeding period and reproduction is their sole purpose in their brief lives.

There are plenty of examples of instinctive sexual behavior from bird life too. The Bowerbird lives in Australia and Papua New Guinea. It builds an elaborate love nest, a bower, as a bird residence, with multiple twigs and scraps it finds in the forest. It then decorates the nest and the entrance with colorful objects like flowers and seashells. When completed it makes mating sounds to lure a female in case the attractive home doesn't do the trick. When a female arrives to check out the prospective mate and paid-for house, the male bird flares its crest behind its head into a bright yellow fan, and dances by stepping sideways back and forth in a unique courtship behavior. It is an exotic mating dance, but doesn't necessarily completely entice the female. If it is early in the mating season the female may want to witness other male displays to see which ones make her more willing. In that case, she abruptly flies away. There is no test to measure the male's despondency at female rejection. But female rejection is a common experience among young males of any species.

The Bowerbird is an exceptional intelligent bird with keen visual acuity, and many claim an artistic bent that qualifies him as an architect erecting models of natural objects with aesthetic touches.

The sexes can replicate themselves in a few different ways. Exhausted salmon swim upstream to where they were born in the river's headwaters where the female mom deposits her eggs in the shallow water. The males then lay their sperm on top of the eggs and the sperm fertilizes the deposited eggs. In this instance, there is no male penetration of the female as there is in most other recognizable species.

Among most insects, mammals, and primates, as we know, the male must insert his phallus into the vagina of the female, eject countless sperm, and let the sperm swim upstream to fertilize awaiting eggs. Strictly speaking, none of this is voluntary. Sex is compelling, more or less automatic once initiated, and an organic imperative. Once reproduction is complete in most species then the organism is expendable. Black widows kill their mates after copulation.

Additionally, there are sexual abnormalities and peculiarities in nature. Goliath groupers that survive along Florida's southwest coast are sometimes born as females but turn into males. They are known as "Protogynous hermaphrodites." No one understands why Goliath groupers have evolved this way.[6] Some barnacles too are hermaphrodites born with both male and female organs. To gain new baby barnacles, however, they use elongated tubes like extended penises that can stretch to several inches to fertilize neighboring barnacles.

And because of research, the advance of medical science and technology, we've learned how to manipulate the process of human reproduction. British medical researcher Robert Edwards worked for years to discover the precise hormonal conditions for fertilizing a human egg in a petri dish. He was successful in 1978, and Louise Brown was born, the first so-called test tube baby. She was created with a separate donated human egg and sperm then combined in a petri dish and implanted in the uterus of the intended mother. In vitro fertilization has led to the births of more than four million humans, the children of one or more infertile parents.[7]

Medical research can now extract a genome and alter the genetic makeup of the egg to prevent the reoccurrence of some inherited family diseases.[8] This advancement does not prevent humans from acquiring dangerous pathogens like SARS (Severe Acute Respiratory Syndrome) or MERS (Middle East Respiratory Syndrome). People got MERS reputedly and initially from camels, and then it got transferred to humans. It won't be the last time cousin animals gave human primates something they didn't ask for.

Sexual Evolution

Nature does not offer reasons why reproduction, and the sex act that is required for it, is differentiated so that that two organisms are necessary for a new organism to be born. What is clear is that gender differences among humans, as experienced by post-adolescents, are some of the most difficult relationships. Though the instincts driving human sexual desire are powerful, this apparently does not impair the instinct of aggression of men towards women.

According to UN reports in 2013, about 35% of all women worldwide have experienced some form of violence from intimate partners, or from non-partner violence. In selected countries (Australia, Canada, Israel, South Africa and the U.S.) between 40% and 70% of women who have been violated, have resulted in their deaths. Among girls 12-16, 83% have experienced some form of sexual harassment. The degree of violence to women by men—caused by lust, anger, misogyny, occasional paranoia, pathologies, and delusional tendencies and deviancy—is almost universally thought to be sexual in nature, and because women have weaker responses to aggression. Men take their frustrations out, or find encouragement for their lusts, often on porn sites on the Internet.

The dominant human emotions for members of the same species or varietals are love, and its accompanying emotion sexual lust, then fear—of female rejection by males, and of attacks from rival males—jealousy, anger, hate of rejection from females and hate against male rivals, sadness, usually because of the passing of a loved one, happiness, and other secondary emotions. Darwin's *The Expression of the Emotions in Man and Animals*, first published in 1872, was one of the first books to document the phenomena of human emotions and the primal proximity to animal emotions.

A simple emotion of pain for an individual can also be condemned if it is directed at a particular group someone else hates. In this age, and in my personal experience, extremism in religion and politics has risen to new levels. This emotion extols the righteousness of racist, bigoted and undignified arguments about those not in agreement with an ethnic, class, or ideological bent. In an age of so much raw emotion I have almost always discovered that those others whom they might consider as enemies are often just down-to-earth people who enjoy a good meal, conversation, friendship and love their families. But such ideologues are propelled by a belief or set of beliefs that borders on obsession, and this often coopts even meaningful dialogue.

On the other hand, advertisers of electronic and print media prey on responsive emotions—happiness, for example, visually portrayed by happy couples in the outdoors, often with dogs or children, or both—to get purchasers of buy their products. The music and entertainment industries likewise cater to emerging adolescent sexual urges by recruiting the young to sing the kinds of songs and make the kinds of movies that thrill and excite adolescent desires for satisfaction or wish fulfillment.

Moreover, more than a million teenage girls get pregnant annually, with nearly 40% ending in abortion and 14% in miscarriage. Since only about 10% of teenage mothers give up their babies to adoption, most teenage pregnancies result in young mothers raising new babies alone and usually without a father. As a result, the young mothers become early losers in the economic struggle for jobs and careers as the majority fail to complete high school. Poverty for them and the children sadly is inevitable.

INDIVIDUALITY AMONG ADOLESCENTS

Biological individual differences within a species have such a wide variability in appearance and behavior that there may be little perceptible difference between where one species actually begins and another ends among certain animal populations. But psychological individual differences typically refer to all the variations existing within the human species. The obvious differences are visible—age, sex, height, weight, hair and eye color. The unobservable individual differences, not as obvious as physical defects, are academic individual differences first diagnosed at school, like mental aptitude, intelligence, memory span, attentiveness, and character.

There are three key concepts that govern human individual differences. First, genetically inherited characteristics already determine in large measure what a person will be. We do not choose our sex or our parentage, or the individual genes that compose our biological composition. Second, people change based on their genes and interaction with the environment. An advantageous environment leads to better chances for survival, a healthy growth, and a longer life. Third, we cannot control the process of development, maturation, and ageing.

Genes, interaction with the environment, and development are the three themes in a study of individual human differences.

The third major influence on individual differences, after genetics and the interaction of genes with the environment, is development, or the natural and evolutionary process of growth, maturation and aging, as when we get pouches under the eyes and wrinkles under the brow. Development is natural to all organic life as a flower goes from a bud to a blossom, or a child becomes a man or woman. Thus, individual differences is not just an idea by which we can compare differences between individuals, but also within individuals as they progress through various life stages to maturity and biological decline. Losing something in life, like virginity or a soul mate, can be a time for grief, or a useful experience for future maturity. The biggest loss as one ages is the loss of youth, from which there is no recovery.

Developmental changes observed in preschool and early elementary-aged children do not appear to be as traumatic as those of adolescence. But they are profound in the child, as cell division and early environmental experiences set the stage for personality differences. Teachers of children of this age need to provide wholesome and positive experiences to foster physical and mental development. Parents and teachers of adolescents need to provide more than just encouragement and adult role models as instructional support, as youth of this developmental stage experience profound and traumatic changes in physical appearance, identity, relationships, and mental and psychological maturity.

Children and youth find their identity in maleness or femaleness, in gender identity, an identity search that begins at birth. Although only about 2-3% of the American population is homosexual, there is a growing social awareness and tolerance for a person's sexual preference that is linked genetically.

Gender, like race, religion or any other identifiable characteristic, is subject to stereotyping, which assumes the behavior of one individual to be typical of a group. Gender roles are learned early in life, even from the colors for clothes chosen at birth. The rituals become more complicated for marriage, birthing, and domestication. Sex refers to the anatomical and biological differences, and gender to the social implications of sex roles in society. The words we use to describe boys and girls and our social intercourse with them is different depending on gender.

For example, kissing as a form of greeting is generally not an American custom among males. Yet it is common among males in Russia and certain European and Middle Eastern countries, as is holding hands. Acquired cultural traits do not limit masculinity or femininity, but define the acceptable expectations within a culture. We are most uncomfortable when we are unsure of another person's sex because we do not know which cultural set of values to use in a social encounter. I know an elderly German gentleman who bows and clicks his heels when introduced to a mature woman.

Adolescence and Emotional Development

The majority of teenagers do not want nor do they seek adult help, because in their formative brains they believe they have all the information they need to make correct decisions as sensibly as adults. Mature adults know this is not the case. The immature brain, an organ in development, appears to be a bit crazy because different regions and circuits of the brain do not mature at the same rate. This usually translates into adolescent bouts of fear and anxiety, as well as heightened risk-taking. Calm sessions of reasonableness are not the rule, and drugs, either prescribed of taken to ease anxiety, are not the answer either.

About the age of 13 a youth places peers and friends as the most important social contacts, replacing parents as the top choice. It is an inherited trait to move out of the nest and learn to survive on one's own or with a social group. Females experience better interpersonal relationships during this age period than boys. Racial or ethnic differences are not as disparate as are gender differences. The social interrelationship between the young person and adults in general continues to be more strained and conflicted throughout adolescence.

In addition to experiencing the physical changes associated with puberty, the adolescent moves from parent-dominated relationships in childhood, through peer-dominated relationships in early adolescence, to romantic relationships in late adolescence. It is only after adolescence and once a more mature self-identity has been established that the late adolescent's social relationship with adults begins to improve.

Hazing among adolescent males is a ritual in college fraternities and football teams. Watch young lion cubs play and they challenge each other by fighting, clawing, biting. It's all preparation for the survival skills needed in the game of hunting. Young primate and non-primate males engage in these genetic inherited behaviors as they approach and enter maturity. But when high school football players haze younger players and violate them sexually they become indicted for criminal activity, justifiably. We have elevated our social behaviors beyond the mere animal instinct. But such activities are ones we are born with. Human primates need to temper them. The adult human primate has to spend the rest of his or her life modulating animal tendencies.

Drugs, however, are on the increase as antidotes to ennui, anxiety and depression, and possibly the higher cost of prescriptions. About 10% of all Americans, or over 32 million, used marijuana in 2012. An ingestion of heroin, a 113% increase, and LSD, an 89% increase, is profoundly on the rise.[9] Of the 2.5 million emergency room hospital visits for drugs in 2011, half were for illegal drugs and the other half for misused medication.

Fear is an important response to the possible approach or presence of danger. Among other primates and mammals, it is essential to recognize when predators or threats are nearby. Adolescents have a heightened sense of such dangers, just like all primates. Adults have learned to be more discriminating about which are serious threats and which sensory cues are benign.

One response to potential threats is for young men to form gangs that become aggressive to other youth. The grouping and bonding is similar to other animals that form colonies to defend against predators. In the same way young adolescent lions still resident in the pride, and before they are expelled when they grow a large enough mane, will attack rival male lions that seek to intrude on the group.

We don't know ourselves fully when we are immature, and perhaps not that much more as adults. But we are good at deluding ourselves by thinking we are heroes or heroines, a fantasy that masks our vulnerabilities. We hide our fears through a series of defense mechanisms, including lies that protect the Ego, that bastion of personal identity, but also of volatile independence.

We are the inheritors of all DNA that has preceded us and our instincts and impulses are derivative of the animal world. Because we have elevated our human image in our culture, we partially deny who we really are and from whence we have come. Such is my main point about our evolutionary past using the example of sex.

The attraction of young females to the songs of the Beatles and to Elvis Presley in the 1960s, and to a whole string of later musical groups to numerous to mention, captivated a whole generation to a heightened exhilaration that energized and elevated sexual energies. Rhythmic beats accentuated the impulses of sexual responses, and the verbal and frenetic shouts of joy showed how close to the bodily behavioral circuits lyrics and music can be to the highly responsive adolescent. Music, rhythm and repetitive sounds stimulate the primary sexual responses.

The peacock ruffles its feathers to attract mates. Some humans, to garner attention and to entertain, dress like peacocks. How do we describe the poor southern boy from Tupolo, Mississippi—and all subsequent imitators—who gyrated his body, dressed like a circus barker, but who became adored by fans for performing his musical talents in outlandish costumes? Rock bands appear to outdo each other in outrageous performance antics while their screaming instruments blare a recurrent rhythmic beat louder than ears are supposed to tolerate. Even our primate cousins would disapprove of such behavior. And we call them the apes when they just act according to their natural instincts.

All sports function as contests to determine the stronger or better competitor, but without actually weakening the opponent. At the conclusion of a contest the contestants demonstrate their good sporting behavior with chivalrous salutes or hugs.

Today's athletic competitions, either in team or individual sports, are participants in events expressive of primal, aggressive instincts in order to become the dominant or alpha or troop commander male. Premier athletes are our alpha males. Male dominance over other males is required to have access to any and all females for mating. Male dominance, whether shown through brute strength, swiftness, agility, or cunning, or some combination of such traits to both genders, is to insure that the dominant contributing genes for procreation will have better chances for survival for the individual and the group.

As in all clans, tribes, troops or prides, there is usually an alpha male who presides. But by and large, there will always be more followers than leaders. There will always be a few more than are actually chosen leaders will aspire to be a leader. The reverse is also true: many more will want to be followers and to have someone else make group key decisions. This will be true generally in gender roles whereby females defer to perceived male strength.

For a few primates like us who may be traumatized, psychologically damaged, or genetically unsuited to lead, being a follower is a necessity for survival. Demagogues, evangelical preachers, charismatic speakers and others of that ilk can more easily sway certain members of the populace to become participants in their cult group. Alpha males may rule for the wrong reasons—not to protect the group from harm but to feed their expansive egos.

Female competitions too are expressive of the instinct to become the dominant female in the tribe, clan or group, to entice the male to have her as the principal mother of the siblings. Once the hierarchy is established, all members of the group adhere to the accepted social standing, either as leader or as followers. All primates exhibit similar social attributes in their behaviors. Females using cunning to attract males may be more successful than males using cunning to track and capture game or animals.

Male and female competitions, whether among animals or the human primate, are responses to inherited impulses to dominate others in the species, and to prepare to defend the group from rival males and predatory acts from other species. The "Will to Power" expressed by the philosopher Friedrich Nietzsche is an inherited will to find one's place in nature's hierarchy, preferably as leader.[10] But it isn't a philosophical will that is at work; it's a biological will. It's all about the will to reproduce of the genes, the lords of our existence that define and express who we really are.

The philosophical will to power has a correlate in biology. According to a research report in *Evolution and Human Behavior* in 2010, male genetic traits are better designed for contest competition than for other sexual selection mechanisms, such as size, muscularity, strength, and aggression. The manufacture and use of weapons probably helped ancestral primate males win contests directly, and deep voices and facial hair signal dominance more effectively than they increase attractiveness.[11] The first primate traits were to win a contest among the other males, not to attract a female.

The human primate is generally monogamous in relationships, but lifetime monogamy is more rare than it used to be, as roughly half the population divorces and marries more than once. There is widespread disagreement among scientists about the origin of monogamy. Fewer than 5% of the more than 5,000 mammal species mate for life. Among our fellow primates, chimpanzees are promiscuous and gorillas polygamous. Among humans, Muslims can take multiple wives, and polygamous tribes exist in Africa and among certain sects in Utah. Except as inspired by biblical precepts, monogamy is unusual in half of humans and most of the animal kingdom.

The reasons for the choice of mating practices are likely multiple, and include territorial space among males and females, choice of food source, and nurturing practices for the offspring.[12] The wandering eye has its origin in our evolution, as it does for most organisms, as the trait supports the Alpha male, the winner of the male aggression contests, for hoarding all the females for his reproductive rights.

Our other main instinct is for obtaining food for survival, and how human population relates to food is the next topic. After sex, food is the personal and group' principal imperative—to feed oneself and the family, tribe or clan. For increasing numbers of human primates food is not an easy provision to obtain. Population has increased unevenly in parts of the world. Even though food production yields have also increased, there are diminishing returns among the poorer populations of Africa and Asia. The uncertain nature of controlling sex to maintain a global manageable population means that there is a dire long-term prospect for human primates, and a possibly disastrous future for many non-primates, including many of Africa's big animals like elephants and lions.

Population and Food Production: Survival at all Costs

Thomas Malthus was the first westerner to correlate population and food with the long-term prospects for humanity. He wrote in 1798:

> *It may safely be pronounced, therefore, that population, when unchecked, goes on doubling itself every twenty-five years, or increases in a geometric ratio.*[13]

Thomas Malthus's clerical mindset, and his emphasis on virtue and vice in population control, got in the way of his mathematics. Theology cannot be a part of quantification and the arithmetic projections between population increases and food production, as Paul Ehrlich pointed out in his explosive 1968 book, *The Population Bomb.*[14] Increases in food yield and parasite-resistant crops with the aid of technology has slowed and delayed the geometric increase in population. Diseases, floods, wars, natural disasters and famines have reduced the precision of the geometric population increase Malthus and Ehrlich warned about.

At the beginning of the Common Era, in about the year one of the Common Era or AD, there may have been approximately 250 million humans. It took another thousand years for the world's population to double to an estimated 500 million, or a half billion. Add another thousand years to 2000 and there were seven billion, a fourteen-fold increase. This rate of increase over even a short term is unsustainable between the number of people and the resources needed to support food needs.

United Nations' Estimates of World Population

1800	980 million (in Malthus's day)
1900	1.6 billion
1950	2.5 billion
2014	7.3 billion
2050	9.6 billion
2100	11-16 billion

Another source: *Scientific American*, December 2014, 311(6), p. 100.

China's population in 2015, at 1.4 billion, is half what the world's population of 2.5 billion was when I was in high school in 1950. But when I first visited China in 1988, the Green Revolution was underway and China was able to feed its massive population. China's government policy of limiting the number of children to one within a family (except for minority Chinese) diminished the rate of China's population growth rate. Meanwhile,

Africa's population, at about one billion in 2015, will increase to three billion within a generation. Africa, Asia and Latin America will account for most of the world's population growth in the coming decades, and the economic and domestic pressure will be to provide food, water, energy needs and sanitation for the expanding numbers. Africa's fertility estimates have been higher than predicted and will account for the largest population increase over the 21st century. Asia's will decline slightly, and Europe, and North and South America will remain stable.

But hunger affects nearly a billion people. Not enough people have access to nutritious food. Enough food is produced to feed everyone. But storage, transportation and distribution have also to be in place for the food to reach those who need it, and money has to be available for its purchase. The affordability of large populations to pay for food is a serious problem. According to World Bank figures, 27 countries have had less than $700 in per capita income, or less than $2 a day for food.[15] The larger the family, the poorer the amount of food is available, even if they are small farmers. The lack of a poor diet creates health problems, reducing productivity and incomes of families further.

By 2050 the earth will need to feed another two billion people. To meet that need, large corporations are moving into Africa, the least food productive continent, to increase yields to feed the planet. As food prices for corn, soybeans, wheat and rice have got corporations scrambling for space, African acreage is cheap and small farmers can easily be pushed off the land by compliant governments.[16] The push to increase food yields on large farms funded by large corporations creates additional problems for small farmers. Such a condition already exists in the developed world where large food corporations control the growing and cultivation of crops and livestock, and the marketing and pricing of food items through large service center grocery chains. Only four corporations control the entire beef production market in the U.S.

War has impacted population growth. There were 37 million civilian and military casualties in World War I. Over 60 million were killed in World War II, the deadliest conflict in human history, reducing the world's population by 2.5%. Wars are always predictable as long as the human primate exists, but their timing and duration are not. A simple arithmetic progression between projected population increases and food production adequate to sustain that population, therefore, is insufficient for accurate predictions about the extinction and/or starvation of the species, as we learned from the simple projections of Malthus.

It is generally accepted that the population decreases after World War II had nothing to do with food at all. Rather, because of the devastation of the war, many people everywhere for the next generation were reluctant to bring children into the world after seeing what so-called civilized people did to each other. What is more predictable is that those with resources will fight or go to war with those with inadequate resources to gain possession of enough natural products, including food, to preserve itself.

> *When, however, population increases, as it has never yet failed to do when the increase of industry and the means of subsistence made room for it, the demand for most of the productions of the earth, and particularly for food, increases in a corresponding proportion.*[17]

Procreation has its delights and disadvantages. It is personally enjoyable, presumably for all mammals. But if every family has several children, as occurs in most Islamic and poorer communities, the lot of humanity is strained. The food breadbasket has its limits overall, as does the family wallet. We love our spouses, our children and our families. In the future, our primate selves as a varietal will have to collectively restrain some desires and the husbanding of resources to maintain our existence. Natural selection also has limitations. The ultimate downsize of all this is our varietal primate extinction brought about, not by predation, but our own mismanagement of the resources nature provides.

Take honeybees as an example of how human mismanagement of the environment leads to food problems. Honeybees pollinate over a hundred U.S. crops, and contribute to about 30% of the nation's annual food supply. Our entire food system is intertwined with the fate of bees. Yet bees too are threatened. They are leaving their hives and abandoning them. The sensitivity of bees to changing environmental patterns is an early warning indication of the decline of insect colonies in food production. Another possible culprit is the excessive use of pesticides in agriculture. Bees work assiduously for the survival of the next generation. That lesson should be enough for us too.

In summary, there is a strong correlation between food and sex that is only obvious if we look at the information about human primate population increases, and the availability of food to feed increased numbers. Americans' love of beef, for example, comes at a price: cattle require 90% of the land used for raising livestock (over dairy products like eggs and pork and poultry), and three times as much water and feed for other livestock.[18] Only Argentinians and Australians eat more beef calories than Americans. It now takes more land, water and feed to produce enough calories daily for survival. Population cannot increase absent an available food supply and the transportation system to get food to markets. There are 2.6 million registered truck tractors on U.S. highways daily, transporting 70% of all agricultural products. If you've been driving anywhere on highways you've seen plenty of them.

Malthus' central premise is right. There is a sliding scale when human primate population overcomes available resources. We just don't know all the variables, like the extent of major natural disasters or extensive wars, to know a timeline or to reach reasonable conclusions yet. But it could be within a century or two.

Then add another factor—rising numbers of immigrants fleeing national and regional disasters, to that volatile mix of sex and food. Immigration policies in Europe and America tend to fall into two political categories. One is based on humanitarian concerns, where a segment of the population feels charitable toward the less fortunate. The other main group views an increased number of immigrants into their country or union as a threat to jobs, more crime and disorder, and increased taxes to pay for welfare programs.

For example, in 2010, 47.3 million people living in the EU, or 9.4% of the total EU population, were born outside their resident country. Immigrants are arriving daily into Europe from less developing countries, primarily Africa and the Middle East, seeking a better life. Many Europeans, responding to far right political pressures, are resisting their arrival because they place a drain on social services. The good will toward immigrant policies, once a sign of embracing foreigners, is yielding to antipathy about urban housing ghettoes and the high taxes needed to support the European lifestyle of foreigners from countries experiencing civil discord or having insufficient resources. Migrating boat populations attempting to enter Europe from North Africa is perceived in such a light. And the migrants will not be bringing the sophisticated and technical skills to contribute to the developed economies in Europe or America.

Immigration, and the xenophobia and fear of the foreigner it generates—both naturally genetic responses—are topics for hot government policy debates. But let me suggest that an integral part of the rise of desperate peoples fleeing their native lands is the result of war, loss of employment and famine, partially exacerbated by the lack of natural resources, including food and water. Governmental, corporate and personal mismanagement and pollution of natural resources, corruption, and the natural depletion of nature's bounty have contributed to a declining availability of what human primates need to live. Immigrants then move to enter those countries where they believe they have an improved chance to survive. Globalization is cultural not just economic. Getting to know your new neighbor may mean entering a new cultural experience. We are witnessing in the fear about the immigration debate a direct consequence of the loss of natural resources, and the relationship between sex and the search for food and other survival necessities.

Social Controls

Human primates are not perfect. The have all the faults of both non-human primates and humans. Achilles had his defective heel and his hubris. All individuals hopefully have more virtues than vices. Likewise, society has its good and bad behaviors. It's essential to control the uncontrollable, the Freudian ID, that unconscious, runaway set of impulses that partially define us, our search for instant self-gratification.

Instant celebrity status has been made more convenient through technology. When someone cannot hope to find a friend, that person can find "likes" on social media by posting photos of a child, dog or cat. When you couldn't make it as a journalist, you can write a blog to enhance your status as a writer. When someone needs to feel important, he or she can record a performance on video doing something silly or outrageous on YouTube. If you seek to be a recognized star, you can go on a reality TV show that let's you believe you are an actor. The opportunities to become self-important are technologically available without the need for vetting by experts. In the end, the incessant search for recognition, and the positive reinforcement that comes with it, is a form of narcissistic self-gratification.

The most self-aggrandizing, self-promoting, image-conscious, dismissive of workers, greedy and hungry for money, cutter of safety corners, and antagonistic of taxes that support public programs are corporations. The Impulse and Self-Gratification Society are also corporations whose motives are rarely humane, absent any other virtue, ethic or known positive value, and only focused on money and quarterly earnings.

As one example, social norms about sex and marriage have changed significantly in the last couple of decades. Sexual activity has replaced parenthood as a social habit. About 66% of America's young adults are not planning for parenthood. The statistics are alarming and getting worse: 40% of young mothers are unmarried; 50% of these young unmarried women are under 30 years of age. The number of African American children living in single parent homes is 72%.[19]

There is no stable, long-term parenthood plan for about half of American children. There can be no more urgent social issue in American society—that involves family finances, responsible parental supervision of children and teenagers, and tax dollars that support poverty programs among them—that impacts who we are. Sexual gratification has been let loose, and it has ruptured the social fabric by tearing the bonds of parenthood.

Finally, sexual gratification, adolescent instincts, and the search for food for survival are not the only biological imperatives.

Paul Roberts in his 2014 book, *The Impulse Society, America in the Age of Instant Gratification*, makes the argument that we have become addicted to our self-gratification and narcissistic impulses and have lost perspective on other social and civic virtues.[20] Self-absorption may be a characteristic of some members of society, but he believes self-absorption has become the reflexes of the whole society. He also criticizes the current belief that the ideology of efficiency, that the greatest social good is production at the lowest possible cost, should be our highest social aim. Focused on short-term objectives, we seem incapable of even advocating for long-term economic or social goals.

Where once we had institutions that helped us gain social composure and lessen personal anxieties—families, friends, churches—we now have dysfunctional family members, friends only on social media who may be half a world away, and declining church membership that only offer donuts in the social center and soporific palliatives as solutions to social problems.

It's possible that the perpetual search for self-gratification, based as it is on biological imperatives, has embedded within its instinct a rush to obtain life's pleasures before a kind of Armageddon descends on humanity. Is it possible that we are all sensing that the good times once realized may be coming to an end, and that we should grab our share before it is too late? Some of this unconscious social instinct may be in the air. But I think that in general the move to seek increased self-gratification is an extreme form of liberty, to do as one pleases and to forsake social obligations for personal pleasures.

One of the first research studies to analyze delayed gratification was done by Stanford Professor Walter Mischel in 1970.[21] In the experiment, a child, usually a preschooler aged 4-6, was offered a choice of a marshmallow (sometimes a cookie or pretzel) as a reward—a form of positive reinforcement—or two marshmallows if they waited until the investigator returned to the room about 15 minutes later. In subsequent follow-up studies on the same individuals, Mischel and his researchers found that those who delayed their gratification had improved outcomes in life as measured by higher tests scores and higher educational attainment levels. Only about a third of the participants delayed gratification for the full, allotted time. Delayed gratification, even in preschoolers, means that the very young who have a degree of self-control, and faith in what adults tell them, succeed in later life.

Food is an obsession among television advertisers, along with vehicles and beer. The portrayal of food on TV is always tantalizing and tasty, sponsored by restaurants and fast food franchises to induce the potential customer to its delights. There are no corresponding messages showing the calorie counts, sugar content, or the health dangers of overeating.

Thus, adult obesity rates have doubled in the past few years, but childhood obesity rates have tripled. Televised advertising food and sugar drinks products appear to be ubiquitous. "All you can eat buffets" welcome big eaters. Weight is now the visible sign of the increase in over-eating. As home cooking has fallen off, and advertising for take-out meals and fast food eateries has proliferated, instant food gratification has diminished social encounters, and family conversations around the dinner table. As we are always on-the-go, food has become on-the-go, the calories be damned.

Americans are less physically active and are getting fatter, according to the United Health Foundation. The obesity rate from 2013 to 2014 increased from 27.6% to 29.4% of adults, and this when obesity is related directed to chronic diseases such as diabetes and high blood pressure. Twice as many Americans have diabetes as was reported two decades ago. About 25% reported doing no physical activity at all. All this puts a strain on health care costs.

The Centers for Disease Control periodically releases data on childhood obesity rates from its Behavioral Risk Factor Surveillance Survey. But a report, "Too Fat to Fight," by Mission Readiness, a retired military group dedicated to preparedness, has alarming consequences. Since 1995, the proportion of recruits who failed physical exams because they were overweight rose to 70%. The report claims this threatens national security. Americans have increased their daily caloric intake by 250 to 300 calories, about half of which comes from sugar-sweetened beverages.[22]

But it isn't just recruits. About 12% of active duty service members (16% among the army) were obese based on body/mass indexes. That percentage has risen 61% since 2002.

To what do we attribute this obesity epidemic? Is it merely the constancy of food advertising? As in all social science pursuits, there is rarely just one casual effect. But among my candidates for probable causes is a lack of self-discipline, and a tendency to instant self-gratification and not to delay for further self-satisfaction, or in this case, improved health and possible longer life. Over-eating, drinking excessively, smoking, drinking or texting while driving, not getting enough exercise, all result in increased health care costs for everyone. The U.S. non-governmental health care industry is driven by expensive technologies and hobbled by poorer quality compared with other developed countries. It's that same health care industry that half the country doesn't believe government should be involved in at all. Yet the private health industry contributes more to the imbalance in the economy that any other sector.

Sex and food. Impulsivity and self-gratification. These are dominant biological markers and instincts we inherit in our genes. More recessive are those genes that help us survive personally, and within a group and which we usually choose not to pursue with as much gusto. The survival genes are those in which we seek to live longer and thus avoid, most obviously, long-term use of tobacco, excessive alcohol intake, and repetitive overeating. The old Roman stoic dictum, *ne quid nemis*, "nothing to excess," is still relevant.

The other recessive gene is to help in the survival of the group, however defined. The first group is the family, then the tribe or clan and finally the society or nation. Patriotism is promoted and we rise to the occasion when American lives are threatened. But in my experience and analysis, civic pride and participation in social programs is not as vibrant as it once was, despite the patriotic gestures of the national anthem and pledges to the flag. Elevating the more altruistic genes, and minimizing the personal pleasures, is the loftier endeavor.

THE END GAME

We can't discuss the beginnings, the procreation part of the human primate and subsequent life activities and stages, and "the thousand natural shocks that flesh is heir to," without commenting on the end of life, the twilight years. The inevitable does occur in time, and not often when anticipated. So preparation for death is a life activity, in the same way financial planning is an integral preparation for retirement. Understanding one's personal sense of mortality may not come until later adult years, but its finality, by whatever means, is unremitting and non-negotiable. It can come from accidents, diseases, random violence, one's own negligence, or any kind of natural or un-natural occurrences.

The traditional killers are internal—diabetes, cancer, hypertension, heart attacks, strokes, atherosclerosis, and cardiovascular diseases, most caused by the amount and kind of food we ingest. One third of Americans are obese, and another third is overweight. Hence, more than half of Americans, because of excesses in food amounts and types of food eaten, increase the nation's health care costs. Maintaining good health is difficult but not for lack of knowledge about what is required: keeping weight reasonable by exercising regularly, eating nutritious meals of mostly fruits and vegetables and some protein, eschewing packaged meals (packaged food producers add sugar and/or salt to their usually unpalatable product), and cutting down on sugars, fats and salts.

The best good news is that the elderly in developed societies are living longer, enjoying better health, and have access to more recreational activities that keep them young, and thus improving the quality of lives.[23] Life expectancy among developed countries is approaching 80 years of age, slightly longer for women than for men. Health care measures have improved personal health, as have advanced medical and pharmaceutical inventions.

When you have a nostalgic moment, return to look at photos of yourself when you were a ripening, slender chrysalis. Notice how you have this distinct sense that the altered face and features that were once you are no longer now you in physique and looks, but are still you in conscious personality. You get a better understanding of your mortality when you attend too many memorial services of people you knew well in your age range.

The passage of time takes its inextricable toll when you look as old as a European cathedral and not as well constructed. It alters you just as it changes the weather and nature, and eventually brings you full circle to that place where you were before you were born, and all your atoms fold back into inanimate matter.

The worst bad news is that those reaching senior status into retirement often have not saved enough—or because of some financial disaster—to live reasonably comfortable in the golden years. For far too many, Social Security payments, those earned income credits achieved while working, are a godsend and a lifesaver. But in these times of economic distress they are usually only available for poverty level subsistence. Others may have had insufficient incomes during the working life to have enough funds for the retirement years. Still others may have had catastrophic and expensive health issues, or had disabled children, and other financial issues that curtailed their own resources for retirement savings. Alternatively, seniors can be duped out of their savings or financially victimized by con artists and corporations.

Perhaps you've heard the expression that getting old is not for sissies. It isn't for the unprepared either. You may not live as old Tom Parr, buried in Westminster Abbey in London, who lived to be 152 years old, but your quality of life can certainly be enhanced. Keeping a comfortable waist size, enjoying modest pleasures and a healthy lifestyle, and staying free of major ailments, will make old age tolerable, no matter how many grandchildren live nearby. Your health will be improved if during life you managed to be moderately thrifty in adulthood so you could live comfortably in retirement. If you have accomplished all this things, congratulations, you are in the minority.

Chapter 4

The Religious Imperative

We are constantly puzzled by what we don't understand, either because we don't have the capacity to fathom what we see, like the universe full of stars and galaxies—or to witness, as I did years ago, the birth of one's child—or because we believe there are invisible forces influencing what we do. In either case, we make a leap of faith to belief in one mystery or the other: in a story that needs a transcendental reality with omnipotent deities, or one that needs deductive logic to sustain its scientific narrative of history and the origin of life.

We cannot know for certain but we appear to be the only primates that have leaped from observation and personal experience to belief systems. We believe in entities that are not present in actual experience and have accorded them operational control over our lives. Entire rituals and behaviors have been organized to perpetuate these understandings, based to some degree on ancient documents purportedly prescribed or dictated by one or more deities through a human medium. With the possible exception of Buddhism, all religions are based on the assumption of an afterlife. It is this belief and supposition that vitiates the laws of observable nature and our alliance with other primates and animals.

Science and religion are two separate realms of human experience, like different colors filtered through a prism. Science does not define morality and ethics, nor does religion explain the workings of nature, though some believe that an omnipotent deity controls all natural and human events. We are usually people of religion. It is our cultural inheritance. The cult and devotion we demonstrate to the flag, standing instead of kneeling, is an offshoot of a religious observance, now a national religion. But because we are still evolving as primates, we are discovering much more about nature and how we relate to it, and about how we understand ourselves.

I believe we first need to acknowledge our evolutionary origin. Next, we need to be aware that our behaviors obey the laws of nature. Third, we need to disabuse ourselves that our value and moral judgments are somehow independent of these realities and totally under our complete control.

Let's explore the dynamics of American religious beliefs and the perpetual conflict with science, evolution and natural selection to better understand this religious imperative, or what we as human primates have believed. Cultivated religious beliefs are stamped on our coinage, in the paper currency, and in our hearts and minds, as religion has influenced the judiciary, the economy, politics, and international relations, and is totally pervasive in American cultural, social and political life. Religion is a cultural accretion and not a naturally inherited characteristic. The anthropological acceptance of a supernatural belief, many laced with specific prohibitory codes, has been the principal impairment to accepting our animal ancestry.

America is a religious nation based on the number of believers who claim a religious affiliation. Because the early American settlers were a decidedly religious people, their early governance system was primarily religious. Hence, those who claim the colonies were founded on religious principles are correct. But each of the colonies were ruled strictly according to the religion of its founders—Puritan in Massachusetts, Catholic in Maryland, Anglican in Virginia. Colonies, later states after the Revolutionary War, discriminated against minorities who were not members of the majority religion. During Madison's lifetime Methodists could legally be jailed in Virginia for preaching their version of Protestantism.

But after the ratification of the *U.S. Constitution* in 1791, uniting all the states into a "more perfect union" than the inadequate *Articles of Confederation* (1776-1797), that statement is no longer legally true. The Constitution guarantees individual freedom of worship, but forbids government sponsorship, promotion or encouragement, a guarantee repeatedly confirmed by U.S. Supreme Court decisions. Government is to be neutral regarding religion and cannot be in the business of promoting any denomination or religious interest. This religious neutrality, however, is constantly under threats from individuals and organizations seeking to promote one or another Christian ritual or practice. Former president Ronald Reagan proposed an amendment so the people could have prayer in schools, a proposal that had already been reaffirmed by the U.S. Supreme Court as unconstitutional.

> *Congress shall make no law respecting the establishment of religion, or prohibiting the free exercise thereof.*
> (1st Amendment, U.S. Constitution)

It's questionable whether or not religious belief can solve any of America's social problems, though religious organizations can help with direct assistance. Because so many are religiously inclined, their beliefs intrude on the secular nature of government institutions and programs, including schools, by demanding prayer in schools, for example, or deciding whether biblical accounts should be included in the science curriculum, or whether stone obelisks of the Ten Commandments belong in public buildings. The Oklahoma state legislature passed HB 1330 by a 37 to 9 margin to place the Ten Commandments at the front entrance to the state capitol, proclaiming that Oklahoma was a conservative state with Christian values, though they clearly did not realize that the Ten Commandments has a Hebrew or Jewish scriptural origin and does not appear in the Gospels or Epistles.

Orthodoxy and History

Neither science nor philosophy admits orthodoxy or certitude. Only faith and religion presume to know absolute truth. Science is skeptical of opinions unless tested with rigorous empirical methods and validated repeatedly, a lesson learned since the Enlightenment and its promotion of experimental science. In that transformative era, science checkmated the dominance of theology as the principal intellectual inquiry, an effort that was only partially successful. The effort to bring intellectual rationality to the instruction of the young certainly does not succeed in religious schools where clergy and laity alike prohibited the teaching of evolution and the physiology of sex because both in their view are inherently wicked. About a third of Americans believe humans have existed in their present state since time began. They are less scientifically skeptical than religiously fundamental, and their faith often negates the evidence from science.

Theologians and clerics are expert at tracking down other peoples' sins. The evangelical orthodox in America's South, since they base their theology on the cosmogony of *Genesis* and the Protestant ethic, are likewise compelled to shun those not of their persuasion, and to condemn them as heretics.

The controversy between religion and science is perennial, less a dialogue than a contentious diatribe of one set of beliefs in opposition to the other. This does not diminish the sparring between differing monotheistic faiths about their own set of beliefs and practices. But almost universally, monotheisms are rabidly opposed to science that tends to invalidate religious claims, especially about life origins, and contrary to any of the sacred texts.

Science, on the other hand, disparages transcendental theories, certainly in medicine, without methods for validating them. The Federal Drug Administration, for example, does not allow pharmaceuticals to be released to the general public unless relevant tests have been conducted on the proposed drug for any primary and secondary effects. The general public—though not necessarily Christian Scientists who believe all sickness can be cured by prayer exclusively—would resist efforts to curtail this standard process that protects public health and safety.

In these controversies between science and religion—and especially evolution that challenges with scientific evidence life origins from the Bible—learning from human history is instructive.

The Stone Age did not end because humans ran out of stone. It ended because humans discovered other means of enhancing their lives based on new and more advanced authentications. The Copper and Bronze Ages did not end experiments in metallurgy. As humans learn new methods of surviving and coping, they adapt to new ways of living and continue to experiment, the ultimate key to civilized advancement in any era. Today, new discoveries in transportation, technology, science and medicine challenge current beliefs and confront us with dilemmas, forcing us to choose new methods for improving the quality of living.

Still, this clash of cultural values between religion and science is an integral part of any democracy. A series of U.S. Supreme Court decisions have prohibited creationist accounts of the origin of life in schools because they are drawn chiefly from religious texts. Even though a majority of citizens are religious, the U.S. Supreme Court has unfailingly held that government cannot advance, promote or encourage religion. Our government must remain neutral in order to protect the rights of minority religions and the irreligious.

Arguments persist, however, favoring divine intervention, known as Creationism, or Intelligent Design, as an alternative voice against the teaching of evolution in schools, especially in science classes. The Intelligent Design advocates claim that human organisms are too complex to have evolved randomly without divine guidance. Their numbers within evangelical Christianity and other religious groups make it difficult to avoid challenges to court rulings excluding religious perspectives from the science curriculum. State laws, like those in Tennessee that prohibited the teaching of evolution that brought to trial the substitute science teacher John Thomas Scopes in 1925, were illustrative of the power and influence of evangelical Christianity, and the belief that knowledge of the Bible, and the most extreme of Calvinist ideas and 13th century theology, surpassed all other knowledge over the evidentiary foundations of modern science.

The Information Age presents a wealth of scientific knowledge about what we have discovered about the universe and ourselves. If we don't universally accept evolution, based on all the accumulated evidence supporting it, we may as well discard everything we have learned about the life sciences and remain in a semi-stone age mentality, anchored in narratives from the Bible, much of it based on much older mythical literature.

The idea of Intelligent Design, that idea opposed to evidence about evolution, is ancient and appears in both Christian and non-Christian literature. Cicero, for example, wrote in *On the Nature of the Gods* in about 50 BCE. "The first point then, said Lucilius, does not seem to require any argument to prove it. For when we look up at the heavens and contemplate the heavenly bodies, what can be so plain and obvious as the existence of a supreme divine intelligence by whom all living things are ruled?"[1]

Likewise, Boethius, the Christian author of *The Consolation of Philosophy* (c. 520 CE), and Roman Consul under the Roman Emperor Theodosius, wrote: "I could never believe that events of such regularity are due to the haphazard of chance. In fact, I know that God the Creator watches over His creation. The day will never come that sees me abandon the truth of this belief."[2]

The author of *Ecclesiastes* also had something to say about life and mortality and how similar life and death resembles the life of animals.

> *For the lot of man and of beast is one lot; the one dies as well as the other. Both have the same life-breath, and man has no advantage over the beast; but all is vanity. Both go to the same place; both were made from the dust, and to the dust they both return. Who knows if the life-breath of the children of men goes upward and the life-breath of the beast goes earthward? And I saw that there is nothing better for a man than to rejoice in his work; for this is his lot. Who will let him see what is to come after him?*[3]

Informed citizens understand this rationale from Intelligent Design because it bolsters religious beliefs. But they also know that there is no reasonable doubt about evolution either, any more than they would doubt the existence of prostate or breast cancer. Hence, about half of all citizens hold contradictory views of the origin of human life. They feel compelled to hold onto a cultivated religious persuasion possibly imbibed from infancy, often the faith of their parents, and to acknowledge that science is also persuasive because of the abundance of validated evidence.

For example, in January of 2006, barely a month after Federal Judge Jones ruled in a blistering 139 page decision against the Dover Pennsylvania School District for attempting to teach Intelligent Design in science classes, a group of parents filed a suit against Frazier Mountain High School in Lebec, north of Los Angeles. The Board of Trustees of the school district hastily met in an emergency meeting on New Year's Day and approved a special education teacher's request to teach an elective course on Intelligent Design and Creationism, despite protests from math and science teachers that this action would undermine the science curriculum. Federal courts have repeatedly struck down attempts to insert religious principles in schools, but the Dover ruling also helped erase doubts about the validity of teaching evolution.

The first major U.S. Supreme court ruling that creationism was religious instruction was *Edwards v Aguillard* in 1987. Well-meaning but legally ill-informed parents will continue to try to convince local school boards of their community's strong religious persuasions, and to present opposition to the teaching of evolution. And local school boards will continue to satisfy parents and challenge the teaching of evolution through alternative curriculum decisions.

There will always be a perpetual tension between strong American religious beliefs and the nature of constitutional non-intervention. In my experience the origin of this conflict is the result of poor and inadequate knowledge about the Constitution. American knowledge about science is even less robust.

THEORIES: WHAT'S IN A NAME?

Religious advocates often dismiss evolution as "just a theory," an offhanded and simplistic remark meant to diminish its scientific persuasion, making it appear that evolution is just a hunch. A theory is indeed speculative, a premise from which to conduct experiments. But evidence about its validated authenticity is not just guesswork but an explanation for how organisms work and have worked in the past. Gravity too is "just a theory," but few would question its reasonableness about how it explains a natural law. Evolution and natural selection, however, is more than a theory. It is the rock solid foundation of all biological science.

Scientists, on the other hand, counter with the question of why anyone would propose the existence of multiple deities, like Father, Son, and Holy Spirit, without any evidence beyond a religious or literary text, or an imperial-sponsored degree, like the document from the Council of Nicaea in 325 creating the dogma of Christianity known as the Apostles or Nicene Creed.

Religious believers and evolutionary scientists may walk in differing paths, and some may try to reconcile the differences between faith and science. Either way, all will have to face some challenging mental gymnastics to find some accommodation between the two.

A 2004 Gallup Poll showed that 45 percent of Americans believed in creationism and only 35 percent in evolution. A Pew Forum poll on Religion and Public Life in 2005 reported that two-thirds of Americans said that creationism should be taught alongside evolution in schools. It also reported that 42 percent of respondents believed that "living things have existed in their present form since the beginning of time."[4] These figures have not varied much over the years.

This means that nearly half of all Americans believe in the separate creation and maintenance of the species, and not the evolution of life forms from one species into another (natural selection), or that humans have evolved from former primates. The polling results also reveal something about the imperfect state of science education in America, or the persistence of religious beliefs about human creation that conflict with scientific conclusions.

According to the Gallup Poll, very religious Americans continue to align with the Republican Party, while moderately religious and nonreligious Americans are more likely to identify as Democrats or Independents, a pattern that has remained consistent for years.

Should all scientists simply give up and yield to all the accumulated evidence and say that the pursuit of science has all been a hoax, since such a devotedly held belief repudiates all known scientific discoveries over the past three hundred years? Or that the earth really was created about 6,000+ years ago, and that religious literature is indeed the basis for all truth and no one needs science to discover new truths? This would be like returning to the absolutist religious orthodoxy of the Middle Ages in Europe and a repudiation of the intellectual advancements of the past several hundred years.

Consider the Venus flytrap and Intelligent Design in the same context. The Venus flytrap is a vegetative conundrum—a plant that is a carnivore. It doesn't have any muscles like an organism, so how is it possible for it to squeeze its cup open so quickly to trap its victims? Congenial flowers, through fragrances and other wily enticements, invite passing insects into their quarters so sticky feet can pick up pollen and scatter it. The Venus flytrap, against most civilized rules of vegetative matter, eats its entrapped guests like other animal organisms.

Is the Venus flytrap a missing link between vegetative and animal organisms? Is the Venus flytrap one of those special divine plants designed to puzzle human intellects? Or is its unique development a product of natural selection, enigmatic to humans, that demonstrates the wide diversity of nature?

The Venus flytrap demonstrates nature's method of concealing its true identity so it can literally entrap visitors and consume them. The insect and animal world is filled with such deceptions about intent. But almost always one is a predator of the other. Throughout all primate life it is imperative to know who is a friend and who a potential predator. Uncertainty and wariness is the psychological consequence, a condition visible in the human primate even while walking down the street, or praying for relief from some distress.

The mysteries of nature are no less confounding than the mysteries of religion, as Augustine so eloquently described in the beginning of his *Confessions.* "Tell me of your compassion, O Lord my God, what you are to me. 'Say unto my soul, I am your salvation.' So speak that I may hear ... Hide not your face from me."

One of the biggest questions is the origin of life itself. It's just easier to believe life happened in a simple, uncomplicated story we can all appreciate, in which an authority figure made the human primate and breathed life into us, as Enlil did to create Gilgamesh in the ancient Sumerian story, *The Epic of Gilgamesh*, one of the world's first literary works.

So when does human life begin? Scientifically, we all began in the world of invisible microbes and will likely die because of them—cancer, diabetes, leukemia, Ebola and the collapse of all internal organs. Poetically, of course, "We are such stuff as dreams are made of, and our little life is rounded with a sleep," as Shakespeare reminded us in *The Tempest.*

But beyond dreams, stars and cells, throughout history we have been elevating our status in attitude and language far above our place in the universe, or even among other organisms. Much of this attitude originated from the biblical account of human origins found in *Genesis*, the first book of the Bible. It's time to puncture that puffed-up, image of humans, that we are a little less than the angels, when really we are just a little more evolutionary advanced than chimps, gorillas and baboons.

Even if many have resolved a mental dilemma between religion and science in favor of religion, many still believe there has to be a sort of "middle way," a sense of spirituality to replace the loss of organized religion. Scientists I know who do not accept religious beliefs often proclaim that non-believers must feel empty and disconnected—a misperception of the religious sentiment imposed on those they believe to be irreligious. Just because some do not have a religious belief, they still can feel just as fulfilled and joyous in their life pursuits as religious believers. Self-satisfaction in life and joy in living does not come exclusively from religious belief. Nor do anxiety and depression arise solely from religious guilt.

But over the centuries, orthodox religion also shaped popular culture and the idea of a special human status was rampant throughout literature, as passages from Alexander Pope's *Essay on Man* (1734) illustrates. a

What would this man? Now upward will he soar,
And little less than angel, would be more;
Now looking downwards, just as griev'd appears
To want the strength of bulls, the fur of bears.
Made for his use all creatures if he call,
Say what their use, had he the pow'rs of all? [5]

So a compelling literary canon, in addition to the scriptural writings, reinforces the western religious tradition and lends credence to the power of the religious imperative.

BEFORE *GENESIS*

Plaintive hymns over the centuries in religious literature record distress at failing to understand what God wants from human life, or how people should act in accordance with divine directives. A Babylonian poet, echoing a typical conundrum about human existence, expressed this sentiment thousands of years ago when his chief deity was Shamash, the sun god.

> *Who knows the will of the gods in heaven? Who understands the will of the underworld gods? Where have mortals learned the way of a god? Here who was alive yesterday is dead today. I am appalled at these things; I do not understand their significance.*[6]
>
> (Babylonian poet, c. 1,200 BCE)

The scientific controversy today is about the origin of human life, and is not so much an argument between science and religion as it is contention between the *Book of Genesis* and more ancient literature, prior to the composition of the Bible, that narrate and parallel the same mythical stories. The story of human creation in *Genesis* was copied by Hebrew scribes from the more ancient parts of the Sumerian *Epic of Gilgamesh*,[7] the Babylonian *Enuma Elish*,[8] and Hesiod's *Theogony*.[9] Metaphorically, we can still consider ourselves as emerging from the chaos that existed in Hesiod, before *Genesis* was written, and from clumps of clay, the foundation material also found in *The Epic of Gilgamesh*.[10]

It is not just that creationism based on literary descriptions in *Genesis* cannot stand against the compelling facts from evolution. It is that *Genesis* itself cannot stand as a point of absolute belief when it has so many literary antecedents from which biblical scribes extracted the principal themes, largely about creation and the human origin of the story of Adam and Eve.

Before theology became a formal study there was myth that became encased in ancient Sumerian, early Babylonian, in the Akkadian epic the *Atrahasis* (c. 1,800 BCE), in Egyptian and Greek literature, like that of the savior of mankind, Prometheus. Hebrew scribes copied these and similar mythical stories and incorporated them to form a Hebrew narrative and literary culture. All creation myths have a literature that precedes in time the writing of any book of the Bible. *Genesis* is purloined literature from several more ancient literary traditions, *Exodus* from Egyptian lore, and the *Psalms* and other hymns from Canaanite literature.

The *Epic of Gilgamesh* predates Homer by about 1,500 years, and the writings of the earliest parts of the Bible by approximately the same number of years, making it the oldest literature from any civilization. Readers will discover at least 12 similarities between its narrative and the stories in *Genesis*. It seems as if only the names have changed. Both stories, Gilgamesh and the *Genesis* creation story, recreate the analogy of the themes of loss of innocence, finding a partner to live a fulfilling life, the insidiousness of a snake, the tree of life, the revenge of the gods, and the search for everlasting life, and angels guarding against the re-entry into a garden.

Where can you find these themes: the presence of Chaos, a battle between good and bad angels, quarreling among the gods, the creation of man, the deity finishing his work on the sixth day and resting on the seventh? All of the these narratives are in the Akkadian narratives of the *Enuma Elish*, discovered in the ancient library of Assurbanipal in Nineveh, in present day northern Iraq, that dates from around 1,200 BCE, or a few hundred years prior to the composition of the first stories of the Bible.[11]

Similarly, the story of the brothers Cain and Abel and Jacob and Esau in *Genesis* has literary predecessors: Gilgamesh and Endiku from the *Epic of Gilgamesh*, Osiris and Seth from Egyptian lore, Cambysis and Smerdis from Persian literature, and the brothers Castor and Pollux from the Greek. Descriptions in each of these national narratives consist of the moral of offending god, even if one is unsure why. But in the end the offense is calamitous because in all these stories one of the brothers invariably dies, often at the hand of the other brother. Seth kills Osiris, Camysis kills Smerdis, Cain kills Abel, and Jacob's actions lead to the banishment, and certain death, of Esau.

It's possible to believe these narrative themes are just coincidental. But the more likely explanation is that they are national stories copied from an ancient original source, probably the Egyptian myth of Osiris and Seth. [12] Hebrews knew the Egyptian stories because chief biblical figures like Abraham and Jesus went "down to Egypt," and Moses came out of Egypt. Hebrews knew the Babylonian stories, based on the Sumerian legends, because they spend years in exile in Babylon during the period of exile bringing the myths back from the cradle of civilization to the land of Zion.

The great civilizations of Sumer, Babylon, Canaan, Egypt and Greek preceded the Hebrews in economic, political, religious, artistic, architectural, military and literary accomplishments. Hebrew society, in comparison to surrounding civilizations, was relatively late in cultural development because of its remoteness to natural resources like water, its tribal and village raiding mentality, and because more advanced societies periodically overran it. This occurred despite the repeated claims in *Deuteronomy* that God would cause all cities to fall before the invading Hebrew believers.

Hebrew scribes usurped literary nuggets from all known civilizations and consolidated various mythical strands into its national literature. Religious believers today typically think that no religious or mythical writing preceded the Bible and consider it all history. Hebrew scribes created Yahweh, a distinct god from the singular El or plural Elohim found in *Genesis*, who absorbed all the combined attributes of previous deities in other cultures like Enlil from Sumer, Baal from Canaan, Zeus from Greek, and Aten and Amun from Egypt.

The creationist disgust of belief in evolution, and especially natural selection, stems almost exclusively from an absolutist view of the creation myth found in *Genesis*. In my decades-long evaluation of the evidence, it is not just science that demolishes the biblical story of creation, but previously existing literature only archaeologically unearthed in the last 100 to 150 years. *Genesis* is based on narratives—in fact plagiarized by Hebrew scribes to develop a Hebrew literary culture—from more ancient literary legends and myths.

Ancient Beliefs Revisited

The evolution of religious ideas, just like evolution in the organic world, is a series of ideas and practices accumulated over the centuries. An early lack of contemporary understanding of the natural world, for example, the causes of thunder, lightning and natural calamities, led to a sense of awe and wonder among the residents of Sumer about five thousand years ago. What followed was the presumptive presence of unseen deities who seemingly controlled the natural world, and the erection of temples that embodied divine presences where they could be worshipped and petitioned for favors.

A belief in immortality in ancient Egypt, and in the concept of a non-material soul that survived death, created some of the burial and embalming practices that are used today. Before Christianity, the Egyptian and Greek dead went to the underworld, where they remained. Indeed, in the epistle known as *I Peter*, the scribe notes that Jesus descended into the earth to visit the dead prophets, not ascended into heaven.

> *Christ also suffered once for sins, the righteous for the unrighteous, to bring you to God. He was put to death in the body but made alive in the Spirit. After being made alive he went and made proclamation to the imprisoned spirits—to those who were disobedient long ago when God waited patiently in the days of Noah while the ark was being built.*
>
> (*I Peter*, 3: 18-20)

Christ journeyed after his reputed resurrection to the Underworld to talk to, and supposedly to rescue, "the imprisoned spirits," the souls of the prophets. It is an act found nowhere in the gospel narratives. But going to the Underworld, as Odysseus did to visit his mother, was well accepted in the Greek and Roman world of the 1st century of the Common Era. The unusual part of this narrative, if one assumes that all the gospels and epistles have the correct biographical information, is that, like all heroic figures from Greek legends, Jesus traveled first downward to the Underworld, not upwards. The journey into the earth is only in one epistle, and the Ascension noted only in the *Acts of the Apostles*. The original gospel biographies have nothing about the afterlife of Jesus.

A literature that extolled ritual and sacrifice to the deities arose in ancient Canaan centuries prior to the formation of the Hebrew biblical canon. The literary presence of multiple heroes and semi-divine figures performing exceptional deeds beyond human capacity can be found in Greek literature as early as Homer's *Iliad* written in the ninth century BCE.

The idea of gaining knowledge of the divine through transcendental means became the central tenet of Gnostics about two thousand years ago and permeated early Christian thought. Doctrines that combined systems of religion based on all transcendental ideas gained momentum in early Greece, flourished throughout the Roman Empire, and eventually suffused Christianity. Personal revelations in subsequent eras, as self-proclaimed prophets like Mohammed, Joseph Smith, Mary Baker Eddy and similar figures, proclaimed that they had had personal divine experiences and thus formed their own religions. Unitarianism is a little different as Protestantism goes, and is, as H L. Mencken once described it, simply a denominational mattress for skeptical Christians to fall onto.

Documents of the early Christian movement were a mixture of Jewish, Greek Gnostic and Neo-platonic beliefs, Greek mystery cults, and Egyptian concepts of the hereafter, the kind of religious diversity that flourished in the Roman Empire, until it became obvious that Roman authorities looked disapprovingly on theological differences, like Arianism among Christians that turned into civil disturbances. The condemned priest Arius believed that Jesus was a man who received divine favors, and was not a divine person, an interpretation favored by Isaac Newton.

And yet in summary, according to the Gallup Poll, the proportion of the U.S. population believing in a creationist view of human origins, now at 42%, has wavered little over the past 32 years. Three in four Americans consider the Bible to be God's word, with 28% saying it should be taken literally and 47% saying it is open to interpretation. Long-term biblical literalism has declined somewhat, while a secular perspective on the Bible has become more common. Nevertheless, theological interpretations of the special place of humans as less than angels, or as having grace, or sin, a semi-divine origin, or reinvention in an afterlife, cannot be a substitute for the over-whelming results of empirical science, or of scientific conclusions about the universe.[13] The percentage of those polled saying God was not involved in human origins is slowly rising.

Church Councils and the Birth of Christian Dogma

Let me return momentarily to the early centuries of Christianity, known at the time as the Nazarene movement, to see how it originally acquired the special dogmas and doctrines.

Against contentious theological divisions among Christians in his empire about the nature of Jesus, whether he was human or divine and to what extent, Constantine convened an assembly of bishop delegates in Nicaea across the water from Constantinople to define what Christians were supposed to believe. The year was 325. The intimidated congregation of bishops—the first ever assembled by an Emperor—concluded their hesitant deliberations under duress, not realizing that they held the faith of all future Christians in their hurried votes.

The hastily and ill-conceived document that emerged, the Apostles or Nicene Creed, incorporated the most fashionable philosophical ideas of the 3rd century, laden with Greek Gnosticism, expressing Hebrew ideas of a Messiah, with Greek ideas of divinity, about a Jewish religious reformer. Theology could not get any more complicated than that. The result was torturous terminology that baffled bishops, councils and scores of philosophers, but not later theologians, who boldly asserted that their vaporous interpretations passed the test of belief in incarnate divinity.

Ecclesiastics decided that peace with Rome was preferable to theological contention. Hence, with the insistence of the religious community, clerics sought a unity of dogma in the

Council of Nicaea in 325 that, once established with imperial favor under Constantine, a Christian, made heresy and dissent unfashionable and then dangerous.[14]

The nature of divinity, the belief in one God, and then the acknowledgement of a trinity of the three-person Godhead, declaring Jesus as consubstantial with the Father God (but not, oddly, with the Holy Spirit), the salvation thesis making Jesus' entry into human life necessary, the coming of Jesus to judge all mankind, the belief in a third god, the Holy Spirit, the belief in one church, in the resurrection of the dead and life everlasting—all this was proclaimed by 316 intimidated delegates at the Council of Nicaea in 325.

The two who opposed the proclamation, Theonas of Marmarica and Secundus of Ptolemais were excommunicated and exiled. Delegates used the ideals of resurrection and immortality from Gnosticism, metaphysical terms like "substance" from Greek philosophy,[15] and the idea of everlasting life and a god who judges (Osiris) from Egyptian lore. None of these ideas were anywhere in the scriptures, but came from existing popular views in the 3rd century of the Common Era. Sensibly, the little-known *Council of Ariminum* or Rimini in 359 passed a degree that stood in opposition to the Nicene Creed. The Roman Catholic Church did not recognize this council.[16]

The Council of Constantinople in 381, convened by Theodosius, and the Council of Chalcedon in 451, would re-emphasize a philosophical imposition on biblical reality and proudly proclaim that Jesus had two distinct natures, one divine and human, and both existed in one person. The abstract Greeks words for being, existence, essence, spirit, substance, nature and person would be instrumental forever in determining the faith of billions by linking metaphysical concepts from Gnosticism, Neo-Platonism and Aristotle's philosophy with divine attributes.

All of these combined philosophical and cultural themes, reputedly generated by the Creator God, confounded the laws of physics and biology. This metaphysical philosophy then became translated into theological dogma and tested the limits of reason, nature and even philosophy, but not literature. It sealed the belief of future Christians for all time as unchangeable dogma.

At the time, belief in partly human and partly divine individuals was common. Gilgamesh, the Sumerian hero of the first epic and known work of literature, *The Epic of Gilgamesh*, was one-third man and two-thirds god. Other semi-divines, born of the union of a deity with a human, were Samson, Perseus, Achilles, and Hercules, among others. Readers of the first few centuries of the Common Era would not have been amazed at half-human and half-divine literary beings, nor of sexual unions between the gods and humans. Such unions were readily accepted literary themes.

Thus, as described by the gospel scribe Luke, the union of a God, the Holy Spirit, with a mortal woman Mary would not have seemed unusual to readers in the 1st century. What is also inconsistent is that both Matthew and Luke describe a virgin birth, where indeed Jesus had no biological father, but they also trace in the same narrative Jesus's ancestry through Joseph. Perhaps they didn't have very thorough editors in those days to note these inconsistencies. Or perhaps they really were writing hero-worship legendary fiction, following a well-established Greek literary tradition, instead of creating complicated theology that violated natural laws.

There are abundant stories in Greek mythology where deities like Zeus copulate with humans like Europa producing semi-divine sons and daughters. Sons and daughters of Zeus included, among scores of others: Aphrodite, Artemis, Athena, Dike, Dionysius, Hermes, Pan and Persephone. But making episodes in legendary literature a part of biographical history, and then unerringly orthodox, however, is a stretch that ignores history, borrows unabashedly from legendary literature, and is ultimately illogical. None of this makes sense in the natural world of primates.

Humankind is central to the drama of religion. But the natural world existed billions of years prior to human habitation. It is very religious but highly improbable that a scribe of an ancient document is presumptuous enough to be eyewitness to human origins or divine traits, or to have heard the voice of God in his ear. To place revelation in this sense as the key to understanding God is, in one sense, to make God dependent on human writing as the primary tool of communication.

But once myth and legend were elevated to the same level as actual biography, history and truth, theology was born and monotheisms and dogmatic religious belief forever became ingrained in western culture. Christian doctrine exists still as an impediment to more universal scientific beliefs, and is therefore a hindrance to intellectual participation in the global world community. Except for humans, religion is not a part of the primate world.

Many notable scholars have investigated religion and its tenets, from the belief in a deity whose existence was unknown but believed to be known, but whose behavior purportedly affected human lives, to the consolidation of monotheistic religious orthodoxy and clerics who defined what belief was.

Isaac Newton wrote more about his readings in theology than he did about science. John Locke published *The Reasonableness of Christianity* in 1695. David Hume was a skeptic but had a life-long interest in religion, and wrote *Dialogues Concerning Natural Religion* in 1779.[17] Thomas Paine was criticized and shunned for penning *Age of Reason* in 1794-95, his book repudiating Christianity.[18] William James wrote *The Varieties of Religious Experience* in 1902.[19] James was the first to note that deeply religious believers are often neurotic, and, trained as he was as a physician, that all states of mind are conditioned by neurons. Rabindranath Tagore, winner of the 1913 Nobel Prize in Literature, wrote *The Religion of Man*, based on a lecture series at Oxford in 1930, his vision and expression of the creative spirit.[20]

This highly selective short list is meant to show how a few of the most renowned philosophers, poets, and scientists also ventured into writing about religion because it engaged the activities of so many, and for their own curiosity and intelligence. Whether devoted believers, skeptics, agnostics or disbelievers, multiple scholars have studied religious belief in themselves and others, and analyzed the doctrines, feelings and devotions to this singular human experience believed by a large majority of humankind.

To conclude, the wisdom of the religious clause of the First Amendment, not only just as a quotation, is that government cannot interfere with the practice of religion, nor can it advance or endorse it. This constitutional imperative has not deterred religious advocates who continue to try from impose their religious prerogatives on others. Everyone has a right to believe in whatever they wish. Religious orthodoxy—whether Christian missionary activity, Mormonism, or Islamic religious ideology—is not a license to suborn people to someone else's beliefs. Nor is it a liberty to impose religion in the public sphere where religious neutrality is a constitutional right.

Religious controversies aside, schools need to accentuate the positive: the necessity of teaching validated evidence about evolution as a minimum scientific preparation for youth. We should not lambaste against religious forms of intrusion into schools, but rather promote the curriculum that needs to be taught—evolutionary principles and the evidence accumulated for their credibility and informed minds in a global world.

At the minimum, such a bare-bones biological curriculum, essential for a informed citizenry, should include topics on Darwin's principle ideas and his data sources, and organic origins, cellular evolution, examples of organic evolution, genes and DNA, and the evolution of humans and other hominids like Cro-Magnon and Neanderthals based on studies from molecular biology and discoveries from archaeology. If education is supposed to be preparation for life, the future demands no less.

Religion is a cultural artifact, a statement not meant to denigrate its importance to those who share its varied beliefs, rituals and pieties. However, its importance is not embedded in our genes but in our culture. It existed in ancient civilizations as expressions of wonder and awe at the unknown causes in the natural world—the energizing heat of the sun, the terrifying voices of thunder, the loud clamor and curiosity of thunder, the astonishment of food growing from the earth, the mystery of human regeneration. Religion also provided solace for the bereavement associated with the death of family members and the sufferings of friends.

These and similar puzzlements of nature led to reverence for unknown figures thought to control these events, transformed them into idolized deities, otherworldly figures who appear to hold the reins of powers over natural forces. Temples arose to placate and temper their powers to the benefit of their devotees. In time, certain individuals—Nanak, Moses, Jesus, Mani, Mohammad, Joseph Smith, Mary Baker Eddy, etc. Buddha excepted—claimed to have become the spokesmen or prophets for these deities, and their pronouncements made new demands upon the faithful and created a panorama of religious belief.

The ideology of these diverse religions would become the impetus for militant aggression toward those assumed to be infidels. Ironically, religious belief in the Almighty and the more noble virtues became the source of primate wars against other humans.

Over time the development of specific rituals morphed into cultural rites and many became accepted behaviors associated with religion. Hands folded in prayer, the symbolic washing of the hands and face, the abstinence of certain foods or beverages, the coupling of music and dance with a religious meeting, standard forms of dress and garments, the avoidance behavior of those of other religions are some examples of behaviors that are not necessarily religious but have become associated with particular religious persuasions.

Human rituals are more or less equivalent to the phylogenetic ritualistic behaviors of animals. Animals display their anatomical features like human clerics who wear distinctive religious clothing to show their status among the group. Some animals exhibit specific behaviors that make them more appealing to others—the peacock's tail, the plumage of the Bird of Paradise— just as devoutly religious human primates attempt to impress their chosen divinity with select behaviors or abstinences. All mannerisms are culture-bound but with origins in genetic traits. The rituals of religion can be found in nature but not the beliefs in divinities or transcendental realities. Only human primates have religious beliefs, a cultural trait unlikely to become extinguished any time soon.

Venus Flytraps: Plants Eating Organisms.

Paradise Lost (Glyptotek Museum, Copenhagen).

Chapter 5

THE PHILOSOPHICAL IMPERATIVE

When a man thinks that he is reasoning he is really disputing, just because he cannot define and divide, and so know that of which he is speaking; and he will pursue a merely verbal opposition in the spirit of contention and not of fair discussion. (Plato, *Republic*)[1]

He who has a taste for every sort of knowledge and who is curious to learn . . . may be justly termed a philosopher.

(Plato, *Republic*, 5:475c)

Are you one of those individuals who likes to solve problems with knowledge and experience, who tells truth to reality, who likes art, beauty and critical thinking, who searches for meaning in life, who holds high standards of ethics and morality? If so, you may be one of the few remaining truly virtuous philosophers. Maybe you just do crosswords and not yoga meditation, but there is something universal about all of us who seek to find our place, to locate our destiny, and to discover new meanings in life, whether in words, in music, in the heart, or in the clouds.

Philosophical questions began about 2,500 years ago, both in China under Confucius and his parables in the *Analects*, in India by the authors of the *Bhagavad Gita* and *Upanishads*, and in Mesopotamia and Greece, when philosophers like Pythagoras became asking questions about themselves and the world around them.

Philosophy as a method of inquiry is, at the minimum, more than two and a half millennia old. Ambrose Bierce once described philosophy as a route of many roads leading from nowhere to nothing. Many of philosophy's ideas and techniques, developed mainly by the ancient Greeks using logic and argumentation, have been subordinated these days to evidence from science and technological advances, especially ancient notions about the constitution of matter and the universe. But despite a few redundancies, the use of logic, thinking and beliefs in certain values are still relevant to contemporary use. Human primates are thinking beings, so we are all amateur philosophers.

You can ask yourself a few questions to see if you fall into any philosophical identities.

Do you believe in the "pursuit of happiness" that the *Declaration of Independence* has declared as one of the "self-evident truths"? Then, whether you realize it or not you are an Epicurean, one who believes that happiness is the greatest good, a point Aristotle emphasized in his *Nicomachean Ethics*, written about 350 BCE, and promoted by Jeremy Bentham (1748-1832), who believed that the practice of utilitarianism improved one's happiness.

Do you believe that a spiritual life is the essence of all reality? Then you accept that there is a transcendental world, and that you are a transcendentalist. If so, then you may also believe that God is unconcerned about what happens in the world. This is the Deist philosophy, a belief held by most of the American founding fathers, including Thomas Jefferson and Benjamin Franklin.

If you believe that virtue is the highest ideal, and that people should take their cues for living from nature, and control their desires in conformity with nature's laws, then you agree with Emperor Marcus Aurelius (121-180 CE), and all Roman stoics like Seneca and Epictetus, whose pithy parables predated those purportedly said by Jesus.

If you think that humankind is by nature depraved and in need of redemption, then that is the principle belief of Calvinists, one of the strictest of Protestant beliefs.

If you believe that souls are immortal, that the soul separates from the body at death, that souls can acquire knowledge, and that virtue can be learned, then you could be a follower of Socrates and Plato, and of the ancient Egyptians who buried their dead with commodities they enjoyed during life to help them on their journey to the afterlife. Lucretius, (98-55 BCE) thought the soul died with the body. The Jewish Sadducees, whom Jesus did not castigate in his peregrinations, did not believe in the afterlife. But the Pharisees did, and Jesus universally condemned them.

If you think that is limited by God's will—or by instincts, or even voices—than the Indian belief in karma, or the Christian belief of God's active participation in human affairs, is your main philosophy. But apart from divine intervention, there is controversy about whether or not people even have free will when brain analyses reveal neuronal activity prior to muscular responses.[2] Recent studies have demonstrated that functional MRI results show that unconscious roots of decision-making occur before actual choices, thus bolstering the claims of neuroscientists and evolutionary psychologists. Cognitive psychologists and philosophers, however, cling to the claim that there is a symbiotic relationship, and that there are viable physiological alternatives to actions based on conscious choice.

The philosophical distinction between determinism and free will is not a choice between absolutes. We do consciously choose actions to undertake, but we are also fated to do so based on certain predeterminations of our primate nature. There are exceptions; some psychopaths really are pathological and insane, compelled to satisfy unwholesome innate and/or violent urges. Their genes made them do it, not the devil. Even the courts allow for pleas of insanity to mitigate punishments of those so deluded.

Individual philosophies are one matter. But political philosophies are also paramount in contemporary society. Benjamin Franklin, Thomas Jefferson, James Madison, John Adams, Alexander Hamilton and George Washington read and admired political philosophers like David Hume, John Locke, Adam Smith, and Jean Jacques Rousseau, all sharing humanistic and secular views. These great forerunners of American democracy read and admired European scholars and subsequently built their seminal ideas into our constitutional documents and republican and representative form of government.

I explore in this chapter the philosophical foundations that once enlivened and revived western civilizations from theological torpor, and whose ideas became the bedrock of American democracy. I have designated it as an imperative for an analysis of who we are because we are thinking primates. It is not absolutely essential to know our philosophical heritage to know who we are. But it is informative because so many of the ideas are anchored in our political origins and in personal reflections. Values are a key element in any philosophy.

Let me start with a generalization based on my observations and readings over several decades. America is losing its extraordinary capacity to revive itself, to prosper as it once did through adversity, to solve domestic problems with pragmatic panache and a unity of purpose for the common good, to hold its image among other nations for its democratic luster and sense of justice. The American philosophy is pragmatism, more evident now in business than in government.

This lack of resolution to solve national problems is partly a philosophical dilemma, but certainly a legacy of our history. Americans are experiencing a collective disappointment in their leaders and an unquiet sense of personal identity mixed with anxiety in an era of inferior cultural phenomenon and conservative obstructionism. It takes cooperation and compromise to achieve laudable goals. These days those virtues are at a premium.

We seem not to be following our inherited western pursuit of knowledge but rather becoming more steeped in a destructive ideology of incivility and hostile antagonism. Yet the only way forward is not to lean on the legacy of radical conservative strategy of obstructionism but to rely on the western pursuit of the value of science and intellectual curiosity, much diminished in today's national discourse. Americans romanticize the value of schooling, but in practice their attitudes and behaviors reek of anti-intellectualism.

Let me expand on this in more detail.

Western Philosophical Heritage

There is more in heaven and on earth, Horatio, than is dreamt of in your philosophy.

(Shakespeare, *Hamlet*)

We typically exist in a social community we have elevated to a political status beyond family, tribe and clan. Non-human primates have very elaborate social bonds that exist as tribal communities, where the tribal leader, or Alpha male, makes the decisions for the community, usually whether to flee or fight. Among social scientists, social organization is studied by philosophy, sociology, anthropology and political science. I will elaborate on this social phenomenon of our social interactions that have both genetic and cultural traits in future chapters.

But let's examine our intellectual legacy from a philosophical and political theory perspective since our cultural ideas, our values, and operating democratic principles originate in Europe in the 18th century when writers abandoned theology and philosophy as a method and the pivot for explaining everything.

Thinking about intellectual and value-laden problems and solving them with logic, clarity, and with reasonable satisfaction, used to be the chosen mental method for educated and informed peoples. Apart from attending higher education to acquire skills for an occupation, the objective of education is to form an enlightened citizenry capable of dealing with personal problems and national dilemmas. The purpose of a liberal education under the Greeks and Romans was freedom, or liberation from ignorance. The appropriate method for acquisition of knowledge was philosophy and its handmaiden logic.

This was the cultural and educational model until the second half of the 20th century when serious discussions about life problems degenerated into self-help remedies and pop culture nostrums. We became a nation of T-shirt philosophies and bumper-sticker slogans to guide us through a myriad assortment of dilemmas.

It would be quaint but instructive to have Alexis De Tocqueville return and to write a contemporary tome about democracy in America, as he did in 1838.[3] Or to seek out the Swedish sociologist Gunnar Myrdal to describe our unique race problem.[4] What we can do is reach down deep inside ourselves to examine how we collectively came to this stage in our lives, to look at least at the ideological foundations of our political identity.

Our democracy was formed by educated men whose western-inspired philosophical ideas from the Enlightenment led to the enactment of our primary documents: *The Declaration of Independence*, *The Articles of Confederation* (adopted by the Continental Congress in 1777, and ratified by all thirteen states in 1781), *The Federalist Papers*, and *U. S. Constitution*, adopted in 1787 and ratified by the states in 1791. The fact that a majority of Americans understand these documents poorly, and the western philosophical ideas that underlie them even less, is one of the disappointments of our educational system. Philosophy may be considered abstract, obtuse and unworthy of examination. But its core ideas created Christianity and American democracy.

Humans have the unique ability among all primates to question their existence and how they organize themselves for individual and collective survival through social contracts. The early Greek philosophers like Thales, Anaximander, Heraclitus and Parmenides, and later Plato and Aristotle, proposed a distinction between existence and essence, as well as being and substance, as separate realities to describe nature. Their ideas appear simplistic and uninspiring today. But at the Council of Nicaea in 325 philosophical definitions of reality defined divinity, and how Trinitarian deities relate to each other for Christianity. But the central point about all of them is that they began with curiosity about the nature of humans and the natural universe. Two of these key ideas that persist today were mind and soul.

MIND AND SOUL: THE HEART OF THE PROBLEM

I turned my thoughts toward knowledge; I sought and pursued wisdom and reason. (*Ecclesiastes 7:25*)

Many are familiar with Auguste Rodin's statute of the man with his chin in his hands gazing pensively downwards. Known as The Thinker, we can all identify instinctively with this activity as we do it constantly during waking consciousness, but probably without reflecting on how or how well we do it. Every mature thinker could be thinking about the process of thinking and trying to answer questions like:

- Is mind divisible or indivisible, material or immaterial?
- Does thinking require sensory information to operate?
- Does thinking require memory to function?
- Can the loss of thinking occur because of physical impairments in the brain, like Alzheimer's?
- Are mind and brain two separate realities?

So what is thinking, and what is the mind doing when it does it? For that matter, what is the mind—if mind is really the thing we think with and not the brain—when we engage in this mental activity? One central question we confront is: How do you know when you know something? Do we know something only when it is stored in cellular memory? Is the mind just the agent or process by which we know?

Mind comes from the Latin word *mens* that meant originally only memory, as in the word remind. However, it has come to mean all conscious human mental activities, habits, and functions. Mind thus implies consciousness but also refers to the totality of all human mental capabilities and activities—thinking, recalling, perceiving, willing, feeling and reasoning. Mind refers not to a specific place or specific function within the cranium but to the aggregate of mental abilities, functions and powers. Its most unique aspect is its ability for spoken and written language. Thinking refers to what the mind does; knowing refers to what it knows.

Below are the six primary examples of models of the mind developed over the centuries.

Models Of The Mind	
Processor towards Ideal Forms (Plato)	Adapted Mind through Evolution
Power of Reasoning (Aristotle)	Processor of Information (Information Theory)
Sensations (Locke, Hume)	Cellular Connections (Neuorosciences)

Originally, soul and mind designated separate parts of a human. Today we use these and similar cognitive terms in education and psychology, words like learning, memory, cognition, thinking, knowing and reason, to identify mental processes. But what do these terms actually mean? Do the modern terms mean the same as the ancient words and definitions?

Aristotle called soul "the principle of animal life." Both Democritus and Anaxagoras, pre-Socratic philosophers, believed that soul and mind were one and the same. Christians assumed only one of the souls of the Egyptians, the everlasting one, and dispensed with the idea of a soul described by Greeks as the animating force of all living things. This view was still the prevailing culturally accepted notion until the neurosciences dispensed with both the idea of mind and soul and retained only the concept of brain.

Greeks borrowed the idea of soul from the ancient Egyptians who believed in two souls, the KA, equivalent to our concept of spirit that survived the body's death, and BA, a spirit equivalent to personality that did not survive death. Christianity absorbed these culturally prominent concepts into its belief.

Rene Descartes in the 17th century revived the body/mind dichotomy, believing they were separate entities. He thus reinvigorated the idea of dualism, a belief that distinguished humans as distinct from animals. He viewed animals as governed by material mechanisms exclusively. But such distinctions and divisions are a disservice to operational reality—and to some degree to animals—good only for abstract speculation and philosophical divination, unhelpful about how primate humans actually behave or function.

Descartes' famous dictum *cogito ergo sum*, "I think therefore I am," implies, falsely, that it is possible to derive one's existence only from the mind thinking about it. But in fact, existence precedes thought. His contributions to the influential philosophical methodology of doubt in his 1637 book *A Discourse on Method*—a borrowing from the Greek skeptics—and to mathematics were more significant.[6]

Here are a few of my conclusions about theories of mind:

- Mind does not exist, only brain
- Mind exists in with body, but both are separate entities
- Neither mind nor body exist independently but act together as a single unit
- There are as many minds as there are many intelligences
- The mind is the result of a slow evolutionary process

My conviction is that there is no organ known as mind, an idea first proposed by the Greeks, and enlarged in meaning by Descartes, and since his day the cognitive theorists. On the contrary, there is only a brain that controls all human and primate responses through the nervous system. Mind is our convenient word to describe a form of cellular energy that we call brain that functions through the storage in cranial neurons of traces of chemical energy we categorize as memory. The only laws of matter are those brains create. Thought and consciousness are operations of the neurons through the nervous system, and not separate functions.

Here are some of my other conclusions based on my understanding of research from the neurosciences as outlined in my book, *Advanced Psychology for Teachers*.[7]

1) Brain as cellular energy operates bodily actions that trigger other electrical responses
2) Brain as cellular energy rearranges recognized symbols that stimulate sensations and responses in others
3) The evidence for brain as energy are available in results from technological scans
4) Brain as chemical energy partially regenerates itself daily, but not enough over time to prevent its eventual disintegration as the organism declines
5) Brain is governed by the physical laws of the universe, not its own separate laws, and only as an integral part of the total organism
6) All the electrical traces located in the brain are there from individual experiences

I have been a witness during a brain surgery and I saw blood, blood vessels, and grey brain matter. I saw no mind. If a mind is supposed to be immaterial, how could I see it? If there is a mind or soul attached to every person, the first question is: how can an immaterial substance exist at all, and how can it be attached to the uniqueness of a body? There are no biblical or other creation stories I know of that describe the separate creation of a spirit or soul.

It's impossible to separate who we are from our state of being, our unitary, corporeal existence. But because Greeks invented intellectual inquiry, we are the inheritors of this mental and educational tradition. Philosophy became the queen of all studies—theology was the king for several centuries—and its pursuit was the principal curriculum in schools for hundreds of years. Philosophical ideas flooded intellectual history, as did empirical science, and gave the Renaissance and Enlightenment its impetus. That tradition abandons scholastic ideas and theological quibbling in favor of scientific evidence.

Self-help books about life goals describe how to navigate through jobs, life traumas and relationships. Socrates and his questioning methods examining life are passe, and business, technology and pop culture are in vogue. In a practical way, we have neglected our philosophical roots, especially in education.

Education, in a word, is all business and preparation for occupation skills, and sadly not universally deemed necessarily for life's quality pursuits. Employability is an important goal but not the only one. Like all schooling, education is preparation for life, providing useful skills in speaking, social contacts, writing, common understandings of history and current events, the role of the physical and social sciences, and humanities in functioning as beneficial parents as well as productive employees. Forming intellectually curious adults is a societal educational mission, not just a retrievable course from a college catalog.

Furthermore, the decline of the humanities has been underway for some time in America and throughout the world. Economic and business majors have replaced history and English at nearly all colleges over the past few decades. The result has been a noticeable erosion of clear and direct writing. Writing is not just a technical skill. It is a life skill. The absence of convincing communication comes partly from not engaging in serious reading of non-fiction literature, and that is in those instances where reading occurs at all. A corollary result is the slow decline of the literary canon, the classics, and, to use an old metaphor, the heart and soul of the humanities.

Let's go back about 2,400 years to examine how our intellectual heritage got jump-started, and then follow the tradition selectively through time.

PLATO (427-327 BCE): THE IDEA OF A REPUBLIC AND THE GOOD LIFE

> *Until philosophers are kings, or the kings and princes of this world have the spirit and power of philosophy, and political greatness and wisdom meet in one . . . cities will never have rest from their evils— no, nor the human race, as I believe—and then only will this our State have a possibility of life.*
>
> (Plato, *Republic*, 5:473 d)

Plato's *Republic* (c. 420 BCE), and many of his subsequent dialogues, are investigations of human values and virtue, key developments in the early history of philosophy. His ideas on the mind, soul and afterlife clearly influenced the development of Christian doctrine. Since Plato, philosophy has stood as one of the main social sciences exploring the human condition, and the *Republic* as one of the most fruitful and exemplary examples of intellectual usefulness. With Plato, the human primate species makes an Olympic intellectual leap forward.

How does justice contribute to happiness? What is the ideal human community? How do ideas promote human development and the cognitive faculties? These seminal questions are the heartbeat of the *Republic*. Plato uses the epic narrative tradition of a dialogue among questioners to propose his model for the good life. But by posing significant questions, he invites us to participate in our mental exploration of life's chief issues as well, if only in dialogue with ourselves.

Plato's genius is that in the early chapters of the *Republic* he uses straw men with robust but un-thoughtful personalities to propose weak answers to his seemingly superficial questions that are actually loaded with more substantive responses. Their careless answers that appear to be responses to simple questions are later found, by Socrates who is the principal protagonist and questioner, to be meaningful in ways the six speaker-participants had not considered. This of course shows the wisdom of Socrates and the folly of ordinary men. But it also reveals something about the illogical and circuitous route of human discourse that often circumvents the real question and winds up in rhetorical dead-ends, non-sequiturs, and incorrect conclusions—something like the responses of many politicians on TV or in the media.

Plato proposes that the state, whereby philosophers or the elite rule, is speculative and idealistic, a political system greatly to be wished for but unrealistic. Plato experienced in his lifetime the disastrous condition of tyranny after the end of the Peloponnesian War (431-404 BCE), the atrocities of an oligarchy, and the unjust death of his friend Socrates under democracy. This made him skeptical of all political systems. His hope was that wise men would rule justly, and that the greedy, ambitious and plebeian politician would be excluded from holding office. The long line of monarchs, the so-called divine right of kings, theocracy, emperors, and dictators over the subsequent centuries has made us all pine for an ideal state ruled by the wise. Plato's proposal of an ideal state is optimistic, but sadly it has been observed only in literature.

In a subsequent chapter on government I will trace the origins of democracy but also show how even our republic is idealized but has evolved as dysfunctional in many ways.

Plato's analysis of mental feats leads with reason, the faculty that gives us knowledge (*episteme*). Reason, he says, is distinct from conviction or faith, a part of opinion (*doxa*). The third faculty is imagination that leads to creativity in art, music, and poetry. Plato does not elaborate on technical skills as Aristotle does. With Plato, civilization begins its exploration of mental states and conditions, a study that still engages thoughtful individuals and defines philosophy. Plato appears to have understood the primate nature of humans.

> *In all of us, even in good men, there is a lawless, wild beast nature, which peers out in sleep.*
>
> (*Republic*, 9:572 b)

He prefers the mindset of philosopher to that of ordinary craftsmen, though is unclear how philosophy can be supported as an occupation. His depiction of tradesmen like carpenters, cobblers and sailors may appear to be quaint but these are still jobs found throughout the world, and are believed to be less important by the supercilious and effete until their skills are actually needed. Occupations have widened and deepened over the centuries since, but Plato's idea of class differences have not. For him, those who seek knowledge are the real power brokers in society, which is why they should be the leaders. But he is also realist enough to acknowledge that his ideal state is unlikely because it is an ideal under any existing form of government, such as tyranny, oligarchy, or democracy.

He uses forms of government to parallel with true justice and with character differences. Justice, he says, is a part of the inner person, where the true self resides. He is an idealist in the form of government, but a realist about human character. "See how sensitive the citizens become; they chafe impatiently at the least touch of authority, and at length, as you know, they cease to care even for the laws, written or unwritten, they will have no one over them." (*Republic*, 8: 563e)

He repudiates the mass of humanity as having little personal discipline and less ethical or moral behavior. Abundant examples of his disdain at human misbehavior and lack of virtue are in Book 8 but are scattered throughout the *Republic*.

Here is another of his observations from Book 6. "You are aware, I replied, that quick intelligence, memory, sagacity, cleverness and similar qualities, do not often grow together, and that persons who possess them and are at the same time high-spirited and magnanimous are not so constituted by nature as to live orderly and in peaceful and settled manner; they are driven any way by their impulses, and all solid principle goes out of them." (6: 503 c)

But he praises the virtues of the philosopher. "Those who belong to this small class have tasted how sweet and blessed a possession philosophy is, and have also seen enough of the madness of the multitude; and they know that no politician is honest, nor is there any champion of justice at whose side they might fight and be saved." (6:496d) Philosophers, he says, have a "taste for every sort of knowledge and who is curious to learn and is never satisfied."

He is unkind to wealth and the wealthy, and, like a later practicing stoic, sees its influence as corrupting. This idea of praising poverty found its way into the writings of Roman stoics like Epictetus and Seneca, in early Christian writings, in a few of the gospel parables of Jesus, and in the *Acts of the Apostles*.

Plato's *Republic* is a tantalizing display of narrative philosophy, a mixture of ethics, psychology, political science, a treatise on how to live well in a political state. Plato borrows heavily on eastern philosophical traditions, like ideas of an immortal soul, an afterlife, and the transmigration of souls. Plato can be said to be a Christian before that religion was established, a transformed Asian with Hindu roots, but in the end a Greek who is (forgive me) the soul of philosophy.

Aristotle and Natural Philosophy

After Plato, one of the early descriptions of a political state based on philosophy was by Aristotle, a student of Plato in about 350 BCE. He made a strong case for the individual as a part of a community in a state where justice prevails. Today, with our legislatures frequently built around opposing ideologies, few laws get enacted because too many elected officials are married to their oppositional bases, and do not appear to be acting for the common good. Aristotle wrote presciently in the first book of *The Politics*:

> *For man, when perfected, is the best of animals, but, when separated from law and justice, he is the worst of all; But justice is the bond of men in states, for the administration of justice, which is the determination of what is just, is the principle of order in political society.*[8]

Aristotle also claims that an individual has a social instinct, an idea I developed in previous chapters that show our ancestral past as a necessary part of a social unit built around defense from predators. The state is the larger social unit that is also built around defense of the country from attacking enemies. The instinctual premise has not changed from the pack, the pride, or the primate troop leader.

Moreover, Aristotle's discussion of education in the last book of the *Politics* sounds contemporary, firmly advocating for public education, and for a common learning program for all. It is a rationale for a common schooling curriculum, and not for private forms of schooling.

> *Again, for the exercise of any faculty or art a previous training and habituation are required; clearly therefore for the practice of virtue. And since the whole city has one end, it is manifest that education should be one and the same for all, and that it should be public, and not private, not as at present, when every one looks after his own children separately, and gives them separate instruction of the sort which he thinks best; the training in things which are of common interest should be the same for all. Neither must we suppose that any one of the citizens belongs to himself, for they all belong to the state, and are each of them a part of the state, and the care of each part is inseparable from the care of the whole.*[9]

The conservative political trend now, in opposition to Aristotle's ideology, is for privatizing schooling—vouchers, charter schools, home-schooling, and other special interests of the parents, and not necessarily the common good of the whole society. Parents have a responsibility for the education of their children but so does the state. The movement today is a conservative tendency to privatize schooling by diminishing the public role in a democracy, and placing schooling more generously in the hands of parents.

Are adults who happen to be parents the most knowledgeable and informed, the most competent providers of skills needed in a global economy? It is of course arguable whether or not any individual teacher has these qualifications either, but at least teachers have passed some professional vetting to enter the profession.

Despite Aristotle's contributions to science and philosophy, his work appealed greatly to Christians because he wrote that men's greatest happiness existed in contemplation, the closest activity resembling the divine. Christian theologians also relied heavily on similar beliefs like that of creation, and the immortality of the soul that are found in Plato's *Timaeus*.[10]

Aristotle's extensive writings, many completed when he was both a student and teacher in Plato's Academy, contain the seeds of our developing democracy, and in many ways the fundamentals of our humanities culture. His encyclopedic writings included metaphysics and philosophy, the arts and poetry, logic and analytics, rhetoric, ethics, dreams, physics, meteorology, marine biology, the motions and history of animals, on the soul, and a variety of essays on his scientific investigations. Yet only 31 of his approximate 200 works survive.

After he was rediscovered by Arab scholars in the Middle Ages and translated into Latin, he became the sole written textbook for all known knowledge except the Bible, and his writings constituted the core of all study except for theology until the time of Isaac Newton (1643-1727). It's impossible to minimize Aristotle's influence and how thoroughly his writings and ideas dominated medieval higher education and the scholarship of the humanities.

THOMAS HOBBES AND ORIGINAL SELFIES

In the same way that our biology is the consequence of all preceding organisms—literally biological descendants of the original DNA—similarly, our representative democracy has emerged from a constellation of ideas emanating from western political philosophy. One key idea was selfishness, or the original non-photographic selfie.

Political philosopher Thomas Hobbes' (1588-1679) most famous work, *Leviathan* (1651), named after the biblical creature, is a long treatise about the state of humanity and its relationship to the state. [11] In his ideal state a sovereign monarch reigned. Because Hobbes lived through England's devastating civil war between the monarchy and the Parliamentarians, he preferred the monarchy to any political alternative. But he did not favor either the divine right of kings or royal succession. Hobbes' argument is that only a strong central state can keep humanity's self-serving interests in check. He rightly concludes that humanity is essentially selfish and seeks perpetual war. Without a strong state of humanity, people are in a state of civil war, fear and insecurity, traits also existing in other primates.

Indeed, the wealthy and propertied men who drafted and voted on the *U.S. Constitution*, and the earlier compact loosely uniting the liberated colonies, *The Articles of Confederation*, embodied Hobbes' view of human nature—that men were more or less always in a state of war, or at least in tension preliminary to war. The rule of government, they argued, was to check those compulsions and hostile tendencies, and hence to foster a stronger social union and contract, especially over the value of property.

The American constitutional founders frequently questioned the status of democracy itself, because of human ignorance and perfidy, and so maintained elements of an aristocracy. The Constitution originally had members of the U.S. Senate elected by state legislatures (since overturned), and an Electoral College (still in place) choosing the president and not individual voters. They had no faith in popular sovereignty, in "one man, one vote," for everything, only in votes for the House, but not for the Senate, the President or the Supreme Court.

Such lack of faith in popular opinion was certainly inspired by Calvinist and puritanical views of human nature as well of those of Hobbes. The founders romanticized about democratic ideals, but simultaneously feared that unregenerate people would be unable to govern themselves. They clearly did not intend to extend the blessing of individual liberty to women, slaves or non-property owning individuals.

It's worth remembering that the protections of the First Amendment, indeed the first ten amendments, were an afterthought of the Constitution itself, not so much forgotten as not originally intended. Indeed, among the phrases of the *Articles of Confederation*, is a specific prohibition against certain individuals: " . . . the free inhabitants of each of these States, paupers, vagabonds, and fugitives from justice excepted, shall be entitled to all privileges and immunities of free citizens in the several States. Exactly who "paupers and vagabonds" were is not described, but it's clear they were not to be accorded any special privileges, protections or rights.

> *It may be a reflection on human nature that such devices should be necessary to control the abuses of government. But what is government itself, but the greatest of all reflections on human nature?*
>
> (James Madison, *The Federalist Papers*, No. 51)

The history lessons of humanity's cruelty are abundant as those Hobbes discussed. Chattel slavery existed in the U.S. for a few hundred years. There were pogroms in Europe and Russia, genocide in Africa and in Europe in the 20th century, the threat of nuclear annihilation during the Cold War, and more recently fear of retaliation by Islamic fanatics bent on eliminating non-Muslim infidels. Primate aggression is a dynamic component of human behavior.

In some respects, Hobbes didn't go far enough estimating mankind's willingness and ambition to kill to gain power or further a cause. Hobbes readily accepted any form of government to a return to base nature that he viewed as a state of anarchy. Rousseau would in his day welcome the "natural man" to avoid the dangers of civilization that corrupted the original state of humanity. There was no indication that he viewed this exit from human civilized life and re-entry into natural existence as equivalent to an animal life or a Stone Age existence, but in effect that is what it would be.

Hobbes lived during England's civil war that lasted from 1642 to 1651, a conflict that resulted in the beheading of Charles I in 1649.[12] The English civil war between the royalists and parliamentarians resulted in a victory for Oliver Cromwell who exacted revenge on his adversaries, and who turned Northern Ireland into a Protestant enclave. Ireland's dual religions and its bitter effects have been conspicuous for animosity after three and a half centuries. England's civil war between Charles I and the Parliamentarians claimed more British lives than were lost during World War I. The civil war, and Hobbes' exile because of his support for the monarchy, had a profound influence on Hobbes. In fact, *Leviathan* was written in France while the war was ongoing.[13]

Hobbes is the first to base his political philosophy on a social contract. Jean Jacques Rousseau would perform a similar service for political theory in the following century in France.[14] His chief point is that people have to relinquish certain rights in order to secure self-preservation.

Hobbes' view of selfishness and antagonism is that of self-preservation. We adhere to civil laws, he says, to avoid deteriorating into ceaseless conflict.

Here are three relevant quotes about his views on humanity's selfishness from his book, *De Cive*, (On the State).

> *In the first place I set down as a principle, by experience known to all men and denied by none, to wit, that the dispositions of men are naturally such, that except they be restrained through fear or some other coercive power, every man will distrust and dread each other; and as by natural right he may, so by necessity he will be forced to make use of his strength, toward the preservation of himself.*[15]
>
> *But it is easily judged how disagreeable a thing to the preservation either of mankind, or of each single man, a perpetual war is. But it is perpetual in its own nature because in regard of the equality of those that strive, it cannot be ended by victory.*[16]
>
> *Wherefore, to seek peace, where there is any hope of obtaining it, and where there is none, to enquire out for auxiliaries of war, is the dictate of right reason, that is, the law of nature.*[17]

His readers obviously at the time understood the "law of nature" as the law of human nature and not as equally implying to other animals.

Hobbes was dismissive of theology and scholastic philosophy and favored a secular and scientific approach to philosophy and politics, and in geometry a more certain method for understanding. His Introduction to *Leviathan* speaks about the mechanization of the body and the state.

Hobbes says that selfishness is largely "mechanical," what we would call today instinctual, and I prefer to call genetic. Few of our actions are totally voluntary. Though we think of ourselves as having free will, almost all our bodily activities are ongoing without our permission: breathing, the beating of the heart, the functions of the endocrine system. Internal organs operate automatically. Our selfish desires are to maintain those bodily processes and our lives, a trait we have in common with all organic life.

Hobbes, in one of his most famous sentences, writes about conflict and about Americans and how they were: "few, fierce, short-lived, poor, nasty, and deprived of all that pleasure and beauty of life which peace and society are accustomed to provide for them."[18] It was less of an indictment and more of an appraisal, and likely very realistic of colonial America.

He is more succinct, in still another famous sentence, when he writes about the humanity of his day: " . . . and which is worst of all, continuall feare, and danger of violent death: And the life of man, solitary, poore, nasty, brutish, and short."[19] This is obviously a pessimistic prediction for the human primate's future bereft of any altruism, idealism, trust in others and any ameliorating virtues. Hobbes is one of the first philosophers to tell it like it is, to reveal the harshness of survival and the search for security and safety from dangers.

But talk to any contemporary who lived through the wars in Vietnam, Afghanistan, Iraq, Syria, or countries occupied by Japan from 1941 to 1945, or Russia in 1943-1945, or European countries occupied by Germany from 1939 to 1945 and you might find wide agreement with the Hobbesian conclusion about aggressiveness in humans.

Hobbes wrote when logic and philosophy still ruled intellectual discourse. His principal authorities were reason and scripture. I question his definition of reason in the same way Immanuel Kant critiqued pure reason in his book of that name. The fact that Hobbes would use "ecclesiasticall" (sic) in the subtitle to *Leviathan* shows how much he relied upon church law and divine intervention.

Today science rules, what Hobbes called "the knowledge of consequences," and there is an abundance of knowledge unknown in his day. There is no separate human nature in the sense Hobbes, Hume and others of the 18th century described it. There is just nature, the spread of genetic inheritances, DNA, and molecules that define all organic beings. Hume's *A Treatise on Human Nature* (1739), for example, was less about the nature of human existence than about advocacy for sense experiences. It was essentially a work about epistemology, how the mind works. But in that era the work was unique because it departed from the conventional thinking about mental abstractions, and emphasized how our ideas emanated from sensory perceptions of exterior reality, a concept that furthered scientific inquiry and deviated from the stale speculations from scholastic philosophy, and the idea that reason and the philosophy of rationalism defined human existence.

What philosophers called laws were just coincidences and correlations of how organisms responded to sensory stimuli. And most emphatically, there is no divine law, such as that Hobbes described, that corresponds to natural law. Laws presume that someone makes rules about how people should act. The so-called divine law is not an *a priori* rule. It is only a chaotic cultural idea over a long period of time that resulted in the human primate, one among many primate species. Nevertheless, Hobbes uses a corollary with math to describe his meaning of reason.

For Reason in this sense, is nothing but Reckoning (that is Adding and Subtracting) of the Consequences of general names agreed upon, for the marking and signifying of our thoughts; I say marking them, when we reckon by our selves; and signifying when we demonstrate, or approve our reckonings to other men.[20]

Today, we might describe what he proposes as recognition of the symbols of language, recording them in memory, and then repeating them in speech. Hobbes uses deductive logic, the same kind of mental application used in geometry.

Mankind thinks it has liberty to do as it pleases but only has primarily reactions to fear and aggression, love, anxiety and responses to the prodding of others. Civil government is society's response to channeling human responses for the good of all, even though much of it is inconvenient to personal freedoms.

The relationship of an individual to the collective, the state, constitutes many of the writings of the best and most inspired of the West's philosophers. I won't explore all of them, but Hobbes is one of my favorites, as is the next we examine—John Stuart Mill.

JOHN STUART MILL (1806-1873)

The 18th century, especially in France, witnessed an intellectual *tour de force* that re-discovered the power of reason and a personal motivation for achievement. The use of reason had lain partially dormant from the Middle Ages when Thomas Aquinas, borrowing reason from Aristotle, had used it as a new methodology apart from faith for uncovering theological truths. Voltaire, Helvetius, Diderot and Condorcet among others intended to use reason to bring about social reform of the masses. Reason, accompanied by enlightened self-interest, such as proposed by Jeremy Bentham (1748-1832), was intended as a moral guide to behavior founded on an empirical analysis of human nature.[21] What all these enlightenment philosophers had in common was that neither the church nor clerics were going to be a compass guide in any human activity. David Hume's *Dialogues Concerning Natural Religion* (1779) was an expression of finding a kind of religion that was not associated with any known church or faith.

John Stuart Mill expanded Bentham's idea of self-interest, what became known as a philosophy of utilitarianism, to include not just reason but imagination, all senses and feelings, and higher moral and aesthetic sentiments.[22] *Principles of Political Economy*, first published in 1848, is his enduring legacy, after Adam Smith's *Wealth of Nations* (1776), a triumphal study of classic economics. It is a bit dated now, but its essential topics—land, rents, landowners, taxation, capital, the influence of government—are still relevant. I will not elaborate on his full schema but discuss some critiques.

In *Principles of Political Economy* (1848), Mill discusses the philosophy of property at length but without examining why a country like Britain can simply take over all the property of another country like India through invasion and occupation. His argument explained in Book V, No. 15 is that of a paternalistic colonizer. He believes it is more important to bring his view of civilization to those colonized, so that there is an economic advantage to taking over another country. This is the same argument American expansionists used to conquer Native American Indian tribes.

Hypocritically, he notes in an earlier section that one of the advantages of a civilized people is that they ought to be able to form decisions about their interests and not have government do it for them. Apparently, he does not yield the same consideration to those colonized by making them fiefs in their own country, to ask them if it was in their best interests to have Britain bring them its version of civilization, or to govern them at all.

The U.S. imported workers from Africa for plantations, making them slaves in America instead of in their home countries. Hence, America too was a colonial master but in a reverse way of governance. Only the country changed, not the process of enslavement. America has not yet reconciled, and only partially repudiated, this odious part of its propriety and paternalistic historical legacy. Its cultural venom of White supremacy runs deep yet in the American bloodstream.

Mill spent 35 years of his life as a civil servant, an examiner in the government East India Company, so in effect he was married to the state and reluctant to question its colonial acquisitions. Perhaps in his considerations of private property, and as an assistant and then Chief Examiner of the correspondence coming from India, it might never have occurred to him that Indians (or the Irish for that matter) had a prior claim to their own property that had nothing to do with colonial masters. Or perhaps he thought that Indian property was not private at all, and that the rape of a foreign country's assets was just a part of the system of capitalism everyone accepted without question. Principles of political economy should always include the negative ethics of colonial occupation, as Mill's does not.

Likewise, Mill seems never to have completely absorbed the complete independence of America from colonial Britain two generations earlier, and that Americans, and therefore to some degree Native Americans, liberated their "private property" from England's colonial control, and from kings who claimed it falsely and illegally as a British royal prerogative.

In addition to favoring colonialism as evidenced by his silence on such injustices, he is undemocratic and elitist. He believes that the poor should be regulated by the upper classes. " . . . the lot of the poor, in all things which effect them collectively, should be regulated *for* them not *by* them."[23] In another section he called them "the less fortunately circumstanced and comparatively helpless multitude." These were the same kind of people Charles Dickens eulogized and immortalized in his novels, especially in *Oliver Twist* and *David Copperfield.* It may have been symptomatic of the age, spoken like a feudal landowner, but Mill, and England in general, is certainly a part of that class cultural divide. Such an attitude is considered highly paternalistic today, yet obviously welcomed by Mill's readers in the 19th century.

Mill does not discuss the nature of governments, such as whether democracy is preferable to monarchy, or even the different parts of government like the role of the aristocracy, monarchy or the judiciary, but rather the influence of government, by which he meant the laws about such topics as rents, imposts and taxation.

What philosophies often imply is that nation states are sometimes the slaves of their own history. Philosophers need to rise above their countries to speculate on issues that affect the human primate. But they are often constrained by the cultural imperatives and the parochial ideas that nurtured them in their intellectual development. Hobbes was certainly affected by England's civil war, and Mill by the keen distinction between England's working and privileged classes, a differentiation that extends into contemporary times.

Finally, certain politicians find solace in novels and fiction, often to such a degree that the social message of the novel becomes a political ideology. Thus, fiction, like Ayn Rand's work, *Atlas Shrugged*, can also become an ideological paradigm for political thought and action. I have been bemused and conflicted by the adoption of an operating philosophy for American congressional legislative action, not from historical political theorists but from fiction. The constitutional founders based their philosophical principles on well-respected political theorists such as Hobbes, Rousseau, Adam Smith, David Hume and John Locke.

Thus conscience does make cowards of us all,
And thus the native hue of resolution
Is sicklied o'er with the pale cast of thought,
And enterprises of great pith and moment
With this regard their currents turn awry
And lose the name of action.

(Shakespeare, Hamlet)

Asian and Enlightenment Philosophical Traditions

At the center of your being you have the answer; you know who you are and you know what you want.

(Lao Tzu, c 550 BCE)

Asian philosophical examinations of who we are have been also major contributions to the analysis of the human primate. Equally intellectual as the western explorations, they have been more mystical and contemplative, an inward peek into the human psyche and the meditative life. In this analysis, we explore what is truly a special element in the human primate, the ability to explore the inner working of a mental state.

The Chinese classics the *Tao Te Ching*, the philosophies of Lao Tzu, Mencius, the *Analects* of Confucius, and the Indian sacred texts of the *Upanishads* and the *Bhagavad Gita* epitomized these investigations into the inner life. Their inquisitive journeys let inquiring minds look inward into human depths.

Readers of these classic Asian texts will realize that many of the pithy sayings and parables, especially from ancient Chinese sages, resemble what many think as naïve mottos suitable only for T-shirts and wall posters. But they were expressions of the earliest attempts at understanding the human primate's relations with nature, with others, the divine and the Self, or who we are. Their value is their timeless descriptions by individuals who reflected and considered the human condition deeply.

The *Bhagavad Gita*, the sacred Hindu script dating from the 5th to the 2nd century BCE, contains several passages in its 700 lines of verses about contemplation, secret wisdom, and enlightenment, all of which became an integral part of the later tenets of Gnosticism and Christianity. Here is a sample.

When the mind . . . shall rest unshaken in divine contemplation, then the goal of Yoga is yours. Seers in union with wisdom forsake the rewards of their work, and free from the bonds of birth they go to the abode of salvation.[24]

I see the pertinence of these thoughtful Asian texts as examples of serious thinking about mystical and contemplative ideas, regardless of how they are expressed, or whether or not the idea had been a previous insight. For example, we regard paintings of flowers, trees or still life as no less significant because they are topics that have been previously painted. So it is with our deliberations of sublime ideas about the meaning and ramifications of human existence, notwithstanding those depicted by the satirical movie *Monty Python's The Meaning of Life* in 1983.

The *Analects* of Confucius are readable and the most accessible of the Chinese parables, closest to the parables of the stoics like Seneca (4 BCE to 65 CE) or Epictetus (c. 50-138 CE) whose sayings paralleled those spoken by Jesus.[25]

Here is one Confucian example of the popular laconic and succinct epigrammatic sayings from the *Analects*. Confucius said, "There are three things which the superior man guards against. In youth, when the physical powers are not yet settled, he guards against lust. When he is strong and the physical powers are full of vigor, he guards against quarrelsomeness. When he is old, and the animal powers are decayed, he guards against covetousness."[26]

Ancient Asian writers reflected deeply and philosophically on the human condition and their insights have proven to be remarkably trustworthy over the centuries, testifying to the permanency and similarity of human life everywhere.

Gnosticism and Christian Mysticism

Christianity too absorbed parts of the mysticism of the eastern philosophies. The most important concept for the Greek and Roman stoic was to be the captain of one's soul, and to find enlightenment and truth within. If we substitute the recurring gospel phrase, "Kingdom of Heaven," for "inner freedom" we find a thematic match that parallels stoic and Asian philosophies. Such themes are embodied in the Roman stoic classic *The Meditations of Marcus Aurelius* (121-180 CE). Gospel narrators portray Jesus as eating in the homes of the rich but preaching on behalf of the poor, just like a Greek or Roman stoic who eschewed wealth and luxuries.

Over time there was a crossover between eastern and western philosophies that survived more or less intact in monotheisms, and in Buddhism and Hinduism, and rediscovered in New Age beliefs. Every religion has its mystical side whereby advocates devote a select amount of time in supplication, in contemplation of the divine truths, or in Buddhism to reach a state of nirvana or enlightenment. The traditional method is meditation, though clearly others experiment with such methods as sweat lodges, clairvoyance, visions, healings and similar remedies to get in touch with the divine, either outside or within themselves.

Greek philosophers of the first century of the Common Era favored Gnosticism, as did adherents of the early Christian movement. Gnosticism, from the Greek *gnosis*, to know, was an esoteric literary and transcendental understanding of the divine in human life, partially a borrowing from Plato and ancient Hindu thought. *The Gospel of John* and *Revelations* are testaments to its influence. Gnostics believed that the spirit was good and the domineering life force. Trapped inside everyone was a divine spark that needed to be kindled. For the non-Christian Greeks, this secret spark was supplied by Sophia, the goddess of wisdom.

Nestorius, Patriarch of Constantinople (385-451) believed that Jesus was two persons with two natures, one divine and one human. He was anathematized and exiled by Pope Celestine I, and condemned by two church councils, the Council of Ephesus (431) and the

Council of Chalcedon (451). The church later declared that Jesus was one person with two natures, one divine and one human.

If one combined the ascetic spirit of the stoics with the contemplative search for wisdom and enlightenment, then one became a monk, the quintessential Christian lifestyle that fused two ancient philosophies into Christianity. Monasteries and nunneries emerged as the principal locales for occupations in the Middle Ages.

Early Gnostic philosophers quickly became theologians when they postulated that Jesus was an example of the divine power in earthly beings. It thus became incumbent upon followers to engage in this mystical side of human mental life—that matter was evil and spirit good—and become a part of the divine energy. In Christian dogma, the Holy Spirit symbolizes this divine power. Gnostic understandings of a savior, of grace, of devils in competition for souls, have become so assimilated into Christian belief that it is difficult to disentangle original Gnostic beliefs from Christianity, though Gnosticism's origin is in Hindu beliefs.

Soon, however, in the early Christian centuries Gnosticism became associated with a heresy because ecclesiastical authorities believed its source came from personal visions and not from bishops and presumably learned church officials, where they thought the true power of interpretation of things divine resided. That ecclesiastical view prevailed for over a thousand years, when even as late as the 16th century the Bible could not be translated into the vernacular languages for fear the populace would misunderstand orthodox meanings of scripture. The rise of Protestantism—creating a plethora of Christian beliefs and rituals, or in some cases like Puritanism, the absence of rituals—altered church orthodoxy and changed forever the stranglehold of the church hierarchy's grip on whom could interpret scripture and doctrine.[27]

One philosopher and literary author who overturned ideas of God, mind, soul and thereby most of the values of western philosophical traditions was Nietzsche (1844-1900). He believed that the only life worth living was the will to power. Those who did so would evolve into Overman, or Superman. He claims that Christianity seeks to subdue this urge. The Overman dismisses God and any supernatural help as superstition, and thus frees himself purposefully from false beliefs and hopes beyond this life. Nietzsche is in the tradition of an intuitive philosopher, but unusually provocative. He appealed to novelists like Fyodor Dostoevsky (1821-1881) whose Nietzschean ideas can be found in *The Brothers Karamazov* and *Crime and Punishment*, and in Jean Paul Sartre's *Being and Nothingness*, his novel *Nausea*, and play No Exit. [28]

Finally, as a personal philosophical exercise to understand who we are, consider personal existence. You didn't decide how or when to enter the world, or will to be who you are, or even to understand the manner of your arrival, the fixation of your gender, choice of birth or physical appearance, because all this, with the assistance of your parents, was determined for you. But even they did not decide from a list what you were to be, or even your sex. There is no Great Decider, no unseen manipulator who independently chooses who, or whether, a human primate will enter life to begin its arduous existence and then, just as unexpectedly in its course, exit it. When you die who you are, the consciousness of your existence ceases and returns to that place you were before you were born.

Speculations about this universal and ancient conundrum, the search for meaning in human existence, have puzzled all humanity endlessly. That search has been the foundation of philosophies and religions. Each of us is a part of the great organic continuum, born without a known reason, dying as all organisms do after fulfilling the urge to propagate its DNA.

Henri Bergson (1859-941), a French philosopher coined the phrase *elan vital*, a living force that he said showed itself in all living things, and that accounts for the creative evolution of life. The transformation of philosophy from abstractions and metaphors for a least the past two millennia can now be located in physical reality—the *elan vital* and similar concepts are what now constitutes the DNA, and the mind is essentially what the brain does. Theoretical distinctions about the human primate that previously dominated philosophical thinking have evolved into physical investigations of the brain.

My search for this noble enterprise began when I was about 15 years old when I realized like a thunderbolt that I had a personal identity. That idea overwhelmed me for days. After studying all major philosophies and religions for the next decades, including 10 years of meditation and contemplation as a Jesuit seminarian, I have concluded that all human primates are no more or no less than an extension of the natural world. The evidence is everywhere for all to see. The fact that we resemble in anatomy and features most of the mammalian world is compelling.

You will not find yourself essentially in any artificially created world: not in your ergonomic chairs and acoustical-paneled cubicles at work, your stylized bedroom and living room at home, or your technology-driven comfort cars. Nor will you find yourself completely in the world of the written word. Look to the natural world, your beginning and ending habitat for who you are.

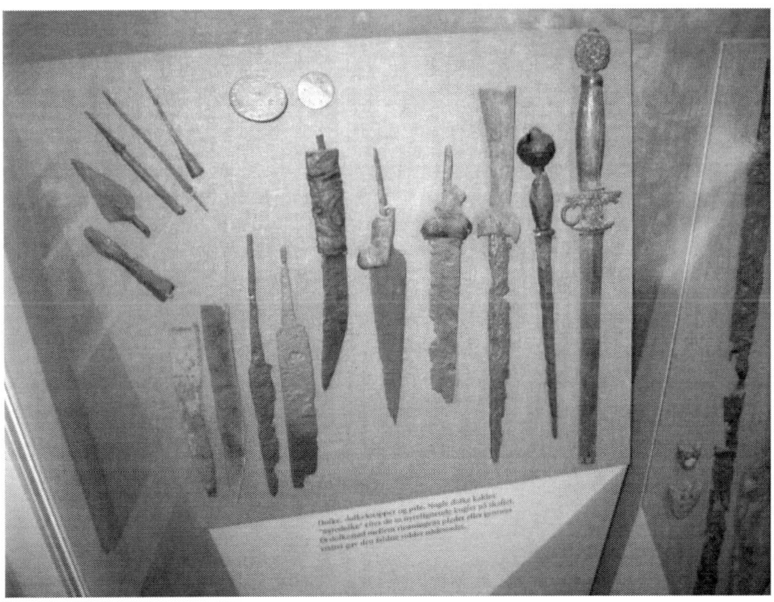

Weapons of Human Aggression (Viking Museum, Ribe, Denmark).

Chapter 6

THE PSYCHOLOGICAL IMPERATIVE

Psychology comes from the Greek word *psyche* meaning soul, although its applied meaning often refers to mind. In Greek mythology, Psyche was the personification of the human soul, and Eros, or Cupid was the Greek god of love who fell in love with her. Psychology is the study of the science of the human personality with all its quirks, emotions, desires, behaviors, and dreams, a probe into who we are, how we think, how we become mature, and how we learn. My objective is less to examine the whole of psychology and more to propose that our study of the psychology of the human primate is partly an extension of the non-human primate, not exactly fortune-cookie science.

Following Darwin's observation, William James (1842-1910) concluded that emotions, cognitions, imagination, perception and similar cognitive abilities evolved as biological adaptations and were not specifically human, but cultural descriptions of select mental behaviors.

Our first and most extensive personal experiences and social bonds are typically within a family environment, and the closest bond is with our mothers, then with fathers and siblings. There are exceptions to these physical and psychological unions, but our earliest behaviors date from these first encounters and can be life-forming. Who we are is not only who we think we are, but how we perceive and act toward others in a social union.

Our deepest psychological adult experiences are usually with a partner, mate or spouse. With whom we choose to live in intimacy, and how we make that delicate choice, is based on how we understand ourselves, and how we view love and partnership in a relationship. Choosing a companion, partner or spouse can lead to happiness, or sometimes to psychological discomfort, depending on our choice and on evolving personalities and extenuating circumstances. Finding a sexual mate is an inborn trait, an imperative in all mammals and primates, and is not psychologically optional, though individuals can choose not to marry or consummate a union.

Throughout our life there often might be multiple partners, some with intimacies, and scores of friends, acquaintances and colleagues that compose the fabric of one's life. If we move from geographic place to place for any reason, we will lose contact with many of those we have befriended, but find new neighbors, friends and work companions. We seek to be nearer and bond more easily with those we find psychologically similar, and to distance ourselves from those we find it difficult to relate to, and this can often be family members. All these behaviors are natural and follow a genetically-driven code.

Our choices are determined by a number of factors, such as age, racial identity, or social class, including level of interest in a social activity, or a physical activity like a sport. We find commonality most earnestly with people we choose to be with principally in racial and social class distinctions. We choose to be familiar with those of our tribe or clan, so to speak.

We have to learn to work with localized fellow workers, but we can choose whom to associate with more often socially, and avoid those we find annoying or disagreeable. The genders will choose each other regardless, determined by genes, their upbringing, and the social venues that allow them to mingle freely. Our social choices are therefore governed first by our perception, and then our emotions toward certain individuals and aversions toward others.

The oldest and most raw emotions are located in the cerebellum at the base of the brain near the brain stem and with the bundle of cords that control the body's nervous system. It is sometimes referred to as the reptilian brain because small mammals and animals have brains that are similar in size and function. Autistic individuals may have smaller cerebellums than the norm.

One of the most social of all animals are dolphins who communicate via echolocation. They swim together, feed together, and gather together even if one of their own is sick and needs attention. Dolphins have high intelligences and highly adaptive social behaviors. Their brain sizes are the largest in the organic world, larger than chimpanzees, our closest primate and ancestral relative. Females reach sexual maturity about nine years and dolphins live to be about 50 years old. These delightful marine creatures are more like us than not.

But besides normal reactions to specific groups or individuals, there exists among us those whose conditions are also conditioned by genetic misalignments. Psychological disorders are also genetically derived and create wide differences in moods and behaviors. We think of these as behaviors as abnormalities. But the origin of these individual differences is in the genes, and their interactive relationship with other genes, or in their omission from a necessary set of genes that should be linked together in a normal sequence. Minor modifications in the sequencing of genes results in individual differences, and in noticeable behavior alterations from the normal.

The National Alliance on Mental Illness notes that one in four adults, about 62 million, live with a mental illness. About 5% live with a serious mental illness such as schizophrenia, bipolar disorder, anxiety disorders such as obsessive-compulsive behaviors, or major depression. The American College Health Association in a 2014 survey found that 54% of college students had "felt overwhelming anxiety" in the past year, a 5% rise from the previous five years.[1] Equally alarming is that about 20% of youth aged 13-18 experience severe mental disorders in a given year.

Since humans are primates, it should not be surprising that other primates have personality disorders, as researchers at the University of Georgia have shown.[2] Other primates have the same neurobiological basis as humans, or vice versa, and hence they are prone to the same kinds of behaviors based on their mental states. Emotionally unstable personality disorders are in all primates in varying degrees. A study of 592 rhesus monkeys discovered that some responded more anxiously than others, and that as many as 30% of their anxiety may be inherited. We are not alone in our moods or emotions.

Twin, family, and adoption studies have shown that around 30% of the between-individual variations are linked to genetics. Several evolutionary mechanisms could explain how genetic variation is maintained in traits, and each can make predictions of rare and common genetic variants that lead to personality variation.

I also want to emphasize that it's difficult to know what is normal, so what psychologists identify as a disorder is a human difference from the average in brain functioning, but not necessarily a negative feature of anyone.

Psychological Disorders Rooted in Physiology

The psychologically maladjusted come in many forms—in mild, moderate or severe stages of a recognized disorder. Individuals with mild or even a moderate psychological disorder can still hold a respectable job, be married and fit comfortable into most social roles. The severely disturbed and pathological are the more palpably recognized and in need of professional help, in some cases institutionalization and imprisonment, to cure them and keep them from harming themselves or others. Suicide is one of the more common results.

> *What the hammer? What the chain?*
> *In what furnace was thy brain?*
> *What the anvil? What dread grasp*
> *Dare its deadly terrors clasp?*
> (William Blake, *The Tiger*)

"Could you please give me a lift to my hair dresser?"

It was my neighbor addressing me one afternoon. I was speaking at that time to another neighbor outside my condo. "Sure," I said to her. "Come over in a couple of minutes and I'll drive you to your appointment."

She was a blond, thin faced with an angular nose and spindly legs, in her early forties. She needed a ride because she had been arrested a week earlier for her second DUI, spent time in jail, and had her driver's license suspended. Over the next couple of weeks I drove her twice to her appointments. During the drives she freely spoke about her personal problems with parents and assorted boyfriends.

She had been seriously injured in a car accident when she was a teenager and two of her friends had been killed in that accident. She had had a serious and life-threatening head injury. Her lawyer father had secured for her a lifetime disability allowance and moved her near the family's second home. Neighbors often heard her screaming into her phone on her back porch.

I went on a domestic trip and when I returned a few weeks later I learned from neighbors that she had committed suicide, overdosing on her medications. I was constantly reminded of her death by the closed blinds on her windows directly across from my kitchen window. She is for me a real-life symbol of how and why personalities are so different, and how brain trauma alters behavior.

We under-estimate the impact that clinical depression can have, especially if compounded with physical disabilities. My experience with a neighbor's suicide elevated the importance of discussing the psychology of who we are, even from the perspective of a disorder.

We cannot peer into a human primate brain when individuals are alive to discover if there are abnormalities. But I am convinced after decades of reading about such research, and conducting my own research studies, that psychological disturbances are the direct consequences of brain anatomy and functions. Every psychological disorder—autism, Asperger's syndrome, bipolar disorder, schizophrenia, bouts of loneliness and depression, among others—all are caused by mild or moderate variations in the DNA or neurons in the brain and affect emotions and behavior, sometimes making us think, if we exhibit such behaviors and symptoms, that we are actors in a bad opera hoping to escape a tragic ending.

Scientific research is constantly identifying causal relations of psychological disorders with genes and brain malfunctions. Williams Syndrome, a rare genetic disorder in individuals who exhibit high levels of trust even in strangers, results from high levels of oxytocin. Medication that blocks oxytocin curbs this excessive and potentially compromising trust.

If troubled by a disorder we think may be our fault, or worse a punishment for misdeeds. It is more than likely a genetic, neurological disturbance that causes it, then we may believe that a divinity like Athena holding back the wrath of Achilles in Homer's *Iliad* will assist us in overcoming our bothersome affliction. But the discomfort, like dyslexia, is within us, and the only consoling contemporary remedies are the limited effects of talk therapy, and prescribed drugs regulating brain imbalances that restore some normalcy to the condition.

Although the cause of Attention Deficit Disorder with Hyperactivity, ADDH, or sometimes ADHD, is unknown, some researchers have discovered that there frequently occurs a flaw in the gene that regulates thyroid hormones in the body. This condition called "thyroid hormone resistance" is often associated with students who are restless and have difficulty focusing their attention, and are often labeled as academically underperforming. Researchers found that in families with a history of this condition that as many as 70% of the children inherited the defective gene that causes ADHD. Ritalin has been the prescribed drug of choice for such children in the past. But the promise now is that there may be a cure for this thyroid hormone resistance disorder through other treatments, and this may reduce the need for prescribed drugs.

One of the more commonly known and recognized abnormalities that affects about four percent of the population is autism. Its characteristics are impaired social interaction or a lack of empathy, even to parents, a lack of eye contact, restricted verbal behavior, a childhood slowness in speech, and repetitive behavior. It is four times as prevalent in boys than in girls, and is unrelated to, and independent of, an individual's intelligence. A related developmental disorder to autism is Asperger's Syndrome.

There is no single genetic misstep to becoming autistic and there are degrees to which it influences the behavior of those who have its symptoms. Like all psychological behaviors that are physiological, there is still more to be learned than exists in the current scientific knowledge base. Autism has only been codified with some definitive success in the past few years. Autistic adults can live very productive and successful lives, as exemplified by Temple Grandin, a Professor of animal science at Colorado State University and author of *The Autistic Brain.* She describes her coping with autism in an era when its effects were imperfectly known and its definition obscure.[3] Many autistic individuals are very sensitive to touches and to sounds. Neuroimaging techniques have revealed other neuroanatomical differences, like longer ventricles in the brain.

Everything understood by the mind can be explained by physical occurrences in the brain and the central nervous system. This novel idea when first iterated was voiced by William

James in the 1890s, Karl Lashley in the 1930s, and Donald Hebb in the 1940s. The present neuroscientific position is that the mental world does not exist independently from the physical activities within the brain, though there is disagreement over how mental integrative functions work. Mental disorders are the result of structural disturbances within the brain. To use an old pun, the brain boggles the mind.

Current research is focusing in on how the brain communicates through sight and through brain waves. Electroencephalograms monitor brain waves and can direct a virtual arm, for example, to manipulate visual objects, thus assisting disabled persons to control movements. But if such advances are common today, consider for a moment how long leeching blood from a sick patient was considered sound medical practice and a proven technique for eliminating disease. As a young boy, my grandmother lathered my chest with mustard plaster in the common belief that it cured colds and extracted fevers, when mustard use was advertised as a universal curing remedy, a slick tool of the mustard industry.

Research Results Linking Genes With Certain Behaviors	
Crime	A Heritability of more than 50% for predisposition
Manic Depression	60-70% for susceptibility
Schizophenia	40-90% heritability
Alcoholism	Twin studies suggest up to 60%
Intelligence	Twin and adoption studies 20-80% heritability
Homosexuality	About 50% heritability

(Source: *Scientific American*, June 1995, p. 125)

Predicting or analyzing human behavior is a bit like trying to forecast an earthquake: there are some seismic indexes that indicate some movement is about to happen, but no one can say for sure when. The laws of physical nature are easier to define and predict than are the rules regulating what or how humans will perform. By whatever name and through whatever discipline or research methodology, we continue the search for who we are and how we think, learn, mature, and behave.

The mind-body distinction is one of the most enduring of this controversial legacy, as I noted in the chapter on philosophy. The most challenging evidence to psychological concepts, especially in the resolution of the mind-body question, comes from both the cognitive sciences and the neurosciences. The neurosciences have shown that there are biological causes to psychological disorders, such as a chromosomal aberration associated with Down syndrome, or a dysfunction in the left hemisphere near Broca's area that causes partial aphasia.

But psychology, like other sciences, is about how facts and ideas that can change thinking about the world. For a psychiatrist or clinical psychologist, the facts need to be combined with a critical thinking process for diagnosing and treating patients with disorders. Failing to do so could result in errors, like not assuming that a child's anxiety is necessarily a sign of abuse, or neglecting a complete medical history in a memory disability patient that might reveal an early history of alcoholism.

THE NEUROSCIENCES: TELLING IT LIKE IT IS

Cognitive neuroscientists study the mind through understanding activities in the brain. The evidence from the cognitive neurosciences lend strong support for the belief that mind is a term from philosophical psychology used for all mental activities, but that all those activities actually occur in the brain and the nervous system.[4] But though we know much more about brain activities, the location of where some originate and how they function, descriptions do not equal satisfactory and convincing explanations of how neural actions translate into love, hate, fear, ambition, or any other psychological trait or behavior.

Evolutionary psychology has emerged as an investigative branch of psychology. Evolutionary psychologists believe, as I do, following Darwin's theory of natural selection, that the brain, like the whole organism, is the result of long evolutionary processes. Whereas the cognitive sciences characterize mental processes as operations or programs, and the cognitive neuroscientists describe the physical components and their interaction in the brain, evolutionary psychologists look at specific evolutionary traits, and comparative traits in other mammals and primates, and not just the cultural aspects of human development.

Evolutionary psychologists, or many social scientists, believe that all humans share a universal, highly organized architecture for responding similarly in certain situations. Primates, from whom they have evolved, share many of these same characteristics, like the search for food, shelter and a mate, nurturing the young and defending the home or nest. They say that human reasoning ability is not a general function of mind, but a diverse collection of evolved mechanisms. In other words, psychological facts and the underlying events in the brain resemble each other in all structural characteristics, as is true in all primates.

In 1985 I began a long-term study of the self-concept of adolescents, seeking to gather evidence about its genetic origins. My theory was contrary to the popular belief then that adolescent self-concept and self-esteem was influenced primarily by environmental factors. Over the next quarter century I collected evidence from a survey form I developed (based and validated on the Tennessee Self-Concept Inventory used exclusively for children) and administered it to adolescents in Cyprus, South Africa, China, Kazakhstan, and Japan, and among Hispanics, Caucasian and American Indian adolescents in America. Based on statistically significant data, I concluded with teams of individuals working in these separate countries, that there was amazing uniformity among adolescents from a wide variety of countries of how they perceive their self-identity, and how they perceive relations with others.

The conclusion was that genetic inheritances were the only plausible explanation for this agreement among diverse countries throughout the world, and not whether they were in a minority culture, or where they lived.[5]

IN SEARCH OF THE BRAIN

Some people say that the heart is the organ with which we think. But it is not so. From the brain and the brain alone, arise our pleasures, joys, laughter and jests, as well as our sorrows, pains, grief, and tears. The brain is the interpreter of consciousness.

(Hippocrates, 460-370 BCE)

Hippocrates, the father of medicine quoted above, suggested in his treatise *On the Sacred Disease* (epilepsy) that the study of physical cures needed to be separated from superstition. He thus placed medicine on a higher level of scientific analysis and above superstition and myth. He is the presumed author of the Hippocratic Oath of trust and honor that a physician takes on behalf of patients. Hippocrates, based on his medical research into suffering and diseased patients—such as that a sword wound to the right side of a soldier's head effected only the left side of his body—was among the first to begin a search for the origin of diseases, many of which we now know are located in the brain.

Neuroscientific evidence now confirms Hippocrates' speculations and, for example, points to the thalmus as the location of schizophrenia. Glandular secretions, especially in the endocrine or hormonal system, such as thyroid, pancreas or sex hormones, can also cause swings in personality and moods.

Similarly, the Greek philosopher Heraclitus who lived in the sixth century BCE, thought that the brain had an enormous space that was infinite. Aristotle (384-322 BCE) believed that the brain didn't have any blood, and the heart controlled everything. The Assyrians believed that the liver was the seat of the spirit or soul. The Egyptians did not think much of the brain, as they sucked it out of the noses of the dead, but for embalming kept other organs they considered vital in the afterlife such as the heart and liver. As Hippocrates suggested, much of what was understood by the brain fell under superstition, myth, and lack of anatomical and scientific knowledge. The modern myths about the brain today are often mechanical analogies: the brain is like a computer, a telephone switching system, or a railroad switching system.

Ancient descriptions of the operations and mechanics of the brain may sound amusing today, but the views held by experts in the past were thought to be accurate and true. Among the various views held were that the brain does not have any blood in it; that the substance of the brain is in a gas; is in a fluid; is in an electrical state; or, today, in biochemical electrical activity.

The oldest con game used to be reading someone's mind. Now, with advanced technology, with special scanners and electronic and magnetic devices peering into the brain and its manifold electrical and chemical operations, reading the mind is literally true. Some of the best evidence of how the brain operates in all its complexity is based on damaged brains. Traumatic brain injury victims, or concussions suffered by football or other contact sport players, usually suffer one of the following:

- A loss of consciousness
- A loss of memory for events just before and after an event
- Some alteration in mental states
- Brain related symptoms such as dizziness or nausea

Head traumas through accidents may temporarily or permanently impair further cell growth and development. Impairments caused by genetic inheritances can produce tumors, aneurysms or other potential brain damage. We are discovering how the whole bodily organism can react to certain allergies in the same way we are learning how the brain controls select behaviors.

Despite our imperfect ability to describe in detail the parts and locations of brain activities, the search is on to locate places that can be targeted so that remedies can be developed to cure specific diseases. The brain is more wonderful than language is capable of describing, or the science of it analyzed. The brain encompasses more than the sum of its varied parts. It is the control center of the entire body and all its functions, the majority of which are automatic, such as the regulation of blood flow, heartbeat, lungs, and the operation of all internal organs.

Most importantly, the brain governs thoughts and emotions and everything we understand by cognitive, affective and psychomotor properties and activities. We know where it is, but not exactly how it operates, or how the reactions of neurons become thoughts.

The brain capacity of a modern adult is about 1250 to 1500 cubic centimeters in size, and weighs about three pounds at maturity. The human brain is the largest primate brain, relative to body weight and proportion. By contrast, the brain capacity of *Australopithicus*, a primitive humanlike primate that existed in Africa about four million years, ago was 400 cubic centimeters. Human primates, through successive organic alterations, have added about 100 cubic centimeters every one million years or so.

> *And still they gaz'd and still the wonder grew,*
> *That one small head could carry all he knew.*
> (Oliver Goldsmith, *The Deserted Village*)

Galen and the Four Humors: Humor Me

Galen (c.130-200 CE) was a second century physician whose anatomical views dominated the next 1,500 years, and were not seriously disputed until the 18th century. He was the first known physician to dissect animal and human brains and served as a physician to a gladiator school where he could observe numerous injuries. He was court physician to Marcus Aurelius, the Roman Emperor, and over 83 of his medical treatises are still extant.

The most significant invention in his day was the engineering feat of aqueducts and conduits carrying water. Not surprisingly, Galen used that analogy to describe the workings of the brain, and proposed that the important parts of the brain were in the fluid-filed cavities, and that arteries carried blood in the brain. Galen's dissections revealed the presence of brain cavities, which he could observe were lined with fluid.

But Galen went further and believed that "states of mind," or personality traits could be deduced from the presence of these fluids, and that an excess of one or more fluids constituted a distinct personality. Since the ancients believed that the universe consisted of four elements, earth, wind, fire and water, then it made sense to believe that the elements that composed the human body were also composed of four elements, fluids, or humors.[6]

The four fluids, or humors that were interpreted as personality traits, Galen proposed were:

- choler, or red for blood (sanguine, or buoyant)
- phlegm, or white for mucus (phlegmatic or sluggish)
- black bile (melancholic)
- yellow bile (temperamental)

Each of these liquids consisted of a different color, and therefore, according to Galen, the relative equilibrium of one or another of these humors determined a person's character and personality. A healthy person had just the right combination of all the humors. Such was the state of psychology regarding personality from the fourth century through Shakespeare's day until the beginnings of modern medicine in the 18th century with surgeons like Paul Broca.

PAUL BROCA (1824-1880)

Paul Broca, was a French surgeon and anatomist, a Professor of Clinical Surgery at the Faculty of Medicine at Paris, and Secretary of the Anthropological Society. Broca believed that certain mental functions occurred in certain regions of the brain. He was the first to point out the subdivisions of the brain in 1851. The traditional view of his day was that mental capacity was determined by brain size.

Broca discovered a specific location for speech, now known as Broca's area. During the autopsy of a man who had lost his speech, and who had selective paralysis, Broca found that there was a large lesion in the left hemisphere where we now know speech is centered. Broca discovered that much of the frontal convolutions had atrophied. The man's brain weighed only 987 grams, or about 25 percent less than the smallest mature human brain. Broca called this condition *aphasia*, or loss of speech.

Broca's cranial contributions intensified the search among neuropsychologists for specific regions where mental functions including speech and writing occurred. The brain was beginning to get mapped out by the late 1880s, and the belief circulated that sections of the brain controlled selective mental activities.

An industrial accident led to another discovery about the psychology of consciousness.

Phineas P. Gage, much admired by his comrades and supervisors, was 25 years old and employed as a construction foreman for the Rutland and Burlington Railroad in Vermont in the summer of 1848. His job was to oversee the laying of new track. In order to lay a straight track line it was necessary to blast away some stone. This consisted in drilling a hole in the rock, filling it with explosive powder, inserting a fuse, then covering the powder with sand by pounding it down with an long iron rod weighing 13 and 1/4 pounds. It was 3 feet 7 inches long, a quarter inch in diameter and had been made to Gage's specifications.

Gage was a blasting expert, but on 4:30 PM this particular afternoon in mid-September he was momentarily distracted by a fellow worker and began pounding the iron bar down into the hole with the explosive charge before the sand has been entered. He struck fire in the rock, there was an explosion, and the iron bar entered Gage's right cheek, went through the base of his skull and exited through the top of his head. He landed over a 100 feet away covered with blood and brain. Phineas Gage had been thrown to the ground but he was awake and conscious.

What follows is one of the most astonishing episodes in brain history, because although Gage would be medically cured within two months, his personality had completely changed. From being the most efficient and capable man employed by the railroad, he became a person unrecognizable by friends and former associates. He no longer had the same character traits and even the railroad would not rehire him, so profound was his personality change. Gage died of an epileptic seizure in San Francisco, after traveling with a vaudeville show as a kind of freak sideshow, on May 21, 1861. He was only 38 years old.

Gage's misfortune revealed that the brain was more than the site of reasoning, but also one of social intercourse and adaptive behavior, and the seat of personal and moral character. Within the three pounds or so of cranial capacity in the normal adult, with its 20-40 billion neuronal connections, have come all the philosophies and symphonies, sciences and knowledge, artistry and architecture civilization knows. The brain has evolved over eons of time, and is the most complex of all known organic structures. Its evolutionary adaptation has allowed it to manipulate the environment for successful human primate survival.

> *The human brain is the most public organ on the face of the earth, open to everything, sending out messages to everything. To be sure, it is hidden away in bone and conducts internal affairs in secrecy, but virtually all the business is the direct result of thinking that has already occurred in other minds.*
>
> (Lewis Thomas, *The Lives of a Cell*)

NEURONS: THE CELLS OF THE BRAIN

Neurons are the most amazing constituents of the brain, and some of the most remarkable elements in all biology. It is the presence of an extremely large number of neurons that make the brain what it is and leads to the complexity of thought. The smallest of the brain cells comprise not only our anatomy but thought processes and emotions.

The number of neurons is thought to be able to contain 100 trillion bits of information. Consisting of a microscopic mass of protoplasm, they are capable of performing all the major functions of what we know as organic life. They operate independently from other life forms. The brain is not an amorphous tissue mass, but a complex and highly organized society of interacting neurons.

Each neuron acts as a sender and receiver of messages to all other neurons by transmitting electrical impulses along a network of connecting circuits. Energy is created by ions, the releasing of protons and electrons from atoms to form new particles. Chemicals exist in the brain as ions in fluids. Impulses are generated by the exchange of positively charged (protons) and negatively charged (electrons) particles flowing through neurons. A semiconductor computer chip acts similarly.

The message the neuron sends first appears as a slight electrical charge traveling from the cell's nucleus outbound along an axon. It finds its way to a dendrite of another cell, the receiving part of a neuron. Separating the axon from the dendrite is a synapse. At the synapse the electrical charge is converted chemically, and then reconverted to an electrical charge as it continues to moves from the dendrite to the nucleus of an adjoining cell, and so on.

One of the more recent discoveries is that the brain has an unusual immune system composed, not of white blood cells that provide guard duty on healthy organs, but of glia cells that provide a similar function in the brain. Glia cells are scattered throughout the brain and are as numerous as neurons. Because these satellite cells destroy invaders, they may also possibly destroy healthy tissue and lead to neuro-degenerative brain diseases such as Alzheimer's disease.

> *The uniformity of the earth's life, more astonishing than its diversity, is accountable by the high probability that we derived, originally, from some single cell.* [7]

Neurons communicate with each other through biochemical and electrical transmissions at a place known as the synapse, or through hormones traveling to a group of cells. The synapse is the junction at which, and through which, cells communicate with each other. It is at the synapse that ions are released that are then transmitted along from cell to cell. Within the molecule lies the gene, the most basic unit of all brain activity and human organic composition.

The genetic programming we inherit in cells can also cause serious diseases and even terminal medical maladies. More than 5,000 human disorders have been attributed to genetic defects. Although cancer patients may feel as if they have been threatened by an alien virus, in fact the malignancy comes from one's own cells as they undergo random changes and mutations. Data about this perplexing dilemma comes, ironically, not from new staining procedures in microscopic investigation but from genealogy and a search into the diseases of ancestors.

It is believed that the process works like this. First, a daughter cell inherits or acquires a defective mutation and passes this defect along to its immediate progeny and all future generations. After two or more generations, some cells acquire enough defects to become malignant cells, and these become cancerous and pass along more abundant defects that cross a thresh hold to cancer producing cells. As I noted above, defective genes can also result in abnormal behaviors and personality disorders.

Select groups of researchers are working on finding the location in the brain of where specific functions occur. In one particular study, researchers tested volunteers who learned the simple motor skill of tapping their fingers against a thumb in a certain numbered sequence of fingers, 4-2-3-1-4. They practiced this sequence daily and underwent periodic testing with a functional magnetic resonance imaging (fMRI) that measures regional brain activity by detecting changes in blood oxygenation levels. Researchers found that more neurons collect in the primary motor cortex for practiced learning than they do for new learning. Select areas, like the primary motor cortex, seem to recruit neurons to meet the demands of the newly learned activity. Moreover, the brain itself is constantly adapting to satisfy behavioral requirements from interactions in specific environments.

Scientists examining the behavior of a man known only as HM, who lost his memory in 1953 via surgery to correct crippling epileptic seizures, gained new insights into the workings of the brain. Scientists believed HM's memory loss resulted from the removal in the brain of six inches of the amygdala and part of the hippocampus, deep in the brain's interior. Thus, the hippocampus and the surrounding cortical regions seem to be the site of where memory filtration occurs. HM could learn a task, like playing a board game, and the next day pick up the game where he left off playing it, but have no recollection of having played it at all.

DNA, the Brain and Memory

All life forms are amazing in their complexity and multiple manifestations. But genes are the most amazing. They are the building blocks of all life, determining eye color, blood type and personality propensities. Everyone is a human fossil carrying the genetic information of every ancestor. Genes are the twisted, intertwining, double-helix forms that constitute the DNA molecule. Genes within the molecule make up a chain of chemical components, mainly adenine, guanine, cytosine, and thymine, often abbreviated as A, G, C, and T. If one of the strands ever breaks apart or separates by uncoiling or unwinding, each separated strand makes another set or template of itself. Genes are self-propagating life forms, the smallest elements in the chain of human shape and behavior, and they literally govern how and what we know and learn and how we retain memories.

The message of how this life-forming information is encoded in the gene is one of nature's best secrets. The message of replication is passed on to the cytoplasm of the cell where it is duly recorded. The Human Genome Project has mapped the order and sequence of each of the genes in the DNA molecule. To illustrate the orders of magnitude, there are about 300,000 working genes in a human being, but over 3 billion DNA letters. As we now know, each human gene has the essential code to develop into a complete human being.

Your DNA is your private museum containing the codes of all your organic ancestors. The microscopic and unseen world of genes and other organic moveable parts is just as awe-inspiring to me, and presumably all of us, as is the visible world of stars and star galaxies, each hundreds of millions years old.

DNA is the abbreviation of the molecule (deoxyribonucleic acid) within which is encoded in the genes all that is necessary to make a complete organism. The human cell contains DNA molecules, which in turn contain the genes that are the programmed building blocks of how the body and brain function and exactly how the whole organism is structured. There are about 5 billion pairs of genes in one DNA molecule, to give you some idea of microscopic size differential.

The age of DNA has recently been pushed back about 80 million years with the discovery and publication of a DNA molecule sequence from 120-135 million years ago, successfully isolated from fossilized plants species and animals. The oldest DNA has been extracted from an extinct species of weevil found in amber in Lebanon and dated to 120-135 million years ago, in an archaeological time period known as the early Cretaceous.

Svante Paabo (1993) reported on his work collecting samples of DNA from ancient human Egyptian mummies. He discovered from analyzing one mummy who lived between 2,310 and 2,550 years ago that DNA molecules survive long after the death of an individual and can be partially reconstructed.[8]

In 1993, Richard Roberts and Philip Sharp were jointly awarded the Nobel Prize in medicine based on their work in the genetic coding. Each separately discovered in 1977 that the genetic coding in DNA was split into segments with gaps in between the genes. In other words, the gene code is not uniform in its messages. Apparently cells can edit their material, skipping over gaps and then splicing some segments together but not others. Scientists had shown that gene therapy can help correct the inherited genes that cause cystic fibrosis, the most commonly inherited fatal disease among Americans, occurring in one of every 2,000-3,000 births.

> *Genetics is the key to the past. Every human gene must have an ancestor...Each gene is a message from our forbearers and together they contain the whole history of human evolution.* [9]

There is an overlapping of genetic material in all organisms. Humans and chimpanzees share 98.4 percent of their genes in common, as I noted above, more than either humans or chimps share with gorillas. It is the complexity of the physical brain and associated human behavior that separate humans from other primates.

In the first study of its kind to demonstrate the location and specificity of a particular gene and its relationship to a personality trait, Ebstein *et al.* (1996) provided evidence that shows a specific genetic locus and its involvement in a normal personality trait, that of novelty-seeking.[10] This finding appears to be independent of ethnicity, sex, or age for subjects who had a higher than average score on a human personality rating scale. There was no significant association found for any of the other personality traits, thus leading to the conclusion that this gene and its location influence this particular personality trait.

There is a growing interest in understanding the evolutionary basis of genetic variation that affects personality traits. According to recent research, around 30% of between-individual variation in personality is because of genetic variation.[11] Although twin, family and adoption studies show positive relationships with personality variables, twin studies alone reveal a higher relationship, often near 70% of shared personality traits. Who we are is mostly determined by what we inherit.

Male and Female Brain Differences

Once we acknowledge the evidence of the evolutionary basis for genetic variation that results in personality traits, the question arises of whether or not there are substantive brain differences among the sexes. The classic contention in the popular culture about stereotypes in personalities—or variances in driving patterns, math scores, intuition, employee pay, sex drives, addictions to shoes or sports, etc.—all by and large emphasize behavioral aspects. The scientific evidence supports a genetic source as the cause of personality and behavioral differences. Males and females do have brains more or less equal in size with minor variations, although men have more grey matter and women more white matter.

Researchers at the University of Pennsylvania reported that male brains have stronger front-to-rear connectivity while those of women are better connected right to left. Women are wired better for socialization, communication and memory, whereas men for coordination, perception and action.[12]

Additionally, there are genetic variations in homosexual and transgender individuals. Researchers at the Stockholm Brain Institute reported that the brains of gay men and lesbian women are similar in brain sizes and in activities of the amygdala.[13]

The brain's gender differences begin in the fetus as sex hormones exert their influence on cranial development. Israeli researchers have discovered that females have language functioning in both hemispheres; men usually have such capabilities just in the left hemisphere.[14]

Perennially, and depending on the observer, there will be either a sense of humor or some indignation at the differences in gender behavior, regardless of what planet, Venus or Mars, we believe these differences originated in the popular and mythological understanding. Now bestowed with an abundance of scientific research into gender brain differences, we must acknowledge that these primate variations have occurred because of a long history of evolutionary modifications, most likely differentiated because of necessary survival mechanisms.

Brain Hemispheres

For years I used as an illustrative aid in graduate educational psychology classes a human skull I had obtained from a state coroner's office. When human skeletons, including skulls, are found for example during road construction, and they are older than a hundred years, they are assumed to be Native American and local tribes are notified. Because tribal members have to rebury the remains in ceremonies, they often do not come to collect them. Skulls can then be loaned out more or less permanently for educational purposes.

I often joked with graduate students that they would feel honored if their skulls could one day be used as instructional aids instead of just lying around somewhere. They were always very attentive when I held the skull as a visual image to discuss the brain and brain hemispheres.

The cerebrum, the Latin word for brain, is divided into two hemispheres, left and right, the directions viewed from the top of the brain in the line of vision. Each hemisphere is a separate organ, and each can function independently of the other, although neither has exactly the same functions. As I noted above, females have more coordination between the hemispheres in language and emotions.

This fact of two brain parts was confirmed in 1844 when A. L. Wigen, an English physician, conducted an autopsy on his lifelong friend who had been normal in all respects, but was found upon examination after death to have only one hemisphere. Wigen reported this in *The Duality of Mind* published in 1844. Even if one hemisphere is partially or even completely destroyed, the other can assume many of the functions of the remaining hemisphere.

The debate about which mental functions like memory are performed by which hemisphere has intensified in recent years. Physiological recovery from a damaged hemisphere is possible because the opposite healthy hemisphere can begin to take on the failed operation from the damaged hemisphere, and new neuronal networks will be arranged. Isolating half of the cortex, or one hemisphere, does not appear to cause any major change in intellectual functioning. What is not so evident is that only one hemisphere wholly assumes a particular mental operation or specialization, such as creativity or logic or mathematics.

The hemispheres are asymmetrical, proportional to each other but not mirror images. Hand dominance is a sign of the brain's asymmetry. In fact, it is agreed that this division of responsibility in each hemisphere actually increases overall intelligence.

Each hemisphere is joined together by a knot of clustered cells known as the corpus callosum. The corpus callosum carries the electrical impulses between the two hemispheres. In extreme cases, the corpus callosum can be severed surgically to curb severe epileptic seizures and prevent death. After such an operation, the patient is unable to communicate between the hemispheres, although is able to function normally in many ways. Each hemisphere receives the visual information coming from the opposite field of vision. Thus, what a person sees on the right side of vision is processed in the left hemisphere, and what is seen on the left is processed in the right.

Vicki, a patient whose corpus callosum had been surgically severed because of epilepsy, could not explain what she saw on her left field of vision in an experiment. But she could draw it perfectly. Why? The picture she saw went to the left hemisphere and is understood there, but the left hemisphere cannot articulate it since pictorial representation is primarily the responsibility of the right hemisphere. Because the corpus callosum was cut, there was no transmission of information between hemispheres. The right hemisphere also understood the picture Vicki saw, but could not verbally respond because language is primarily in the left hemisphere. However, she could draw it.

The right hemisphere processes the information received from the left side of the body, the left hand and the left eye. It "sees" pictures and not words, processes imagination, and "sees" shapes in space. It is able to recognize faces and places. The loss of a certain part of the right hemisphere leads to a loss of facial recognition. The right hemisphere appears to be the principal location of visual and spatial memories.

On the other hand, the right hemisphere seems to be poor at problem-solving and certain logical functions, but gifted at facial recognition and some emotional responses. More significantly, the right hemisphere apparently does not process Western language; that is apparently the function of the left hemisphere. However, the right hemisphere can process ideographic or picture languages, such as Chinese and some Asian languages because these languages rely upon pictorial representation to convey meaning.

Consequently, different cultures, like the Japanese, may process some words in the right hemisphere because individuals "see" a meaning and do not just "hear" a word association. In the same way, gifted musical individuals can see the music while it is being played.

Each hemisphere is divided into four lobes.

- Occipital lobes, located at the back of the brain, control vision and sight
- Temporal lobes, near the temples, control hearing and possibly auditory memory
- Frontal lobes, behind the forehead, are the largest lobes and command most of the brain's activities
- Parietal lobes, after the Latin meaning "wall," are the upper anterior lobe of the cortex probably responsible for word processing

Covering the whole brain is a one-eight inch thick sheath known as the cortex. The cortex, the so-called grey matter of the brain, exists in folds over the cerebrum, whose nerve fibers are the white matter of the brain. Scientists do not yet know with certainty what the function of the cortex is, but it is thought to be responsible for making the decisions about information processed elsewhere in the brain.

From all this research we can conclude without equivocation that the brain is the control center and seat of personality and our psyche, as it is for all primates. We are coming to learn of our unusually long evolutionary past and the three-pound organ that governs all human primate behavior.

In the following section are three animal psychologists whose experiments confirmed how human primate behaviors are so closely allied with other animal and primate behaviors. Ivan Pavlov, Wolfgang Kohler and Karl Lorenz, typified research that showed how alike we are with other mammals and primates.

Let us turn now to the experimental father of behaviorism, Pavlov, whose studies in behavioral response in animals set the groundwork for the classical theories of behaviorism in the early 20th century.

IVAN PETROVICH PAVLOV (1849-1936)

Rules governing behavior arose initially from the work of the Russian physiological experimenter, Ivan Pavlov, whose name is now associated with classical behavioral conditioning. This experimental study of physiology and organic behavior became the basis for psychological investigations. Behavioral science is based on the physiological fact that organisms operate as a unified whole and not as separate parts acting independently.

Pavlov was born September 14, 1849, in Ryazan, a town about 150 miles south of Moscow the son of a village priest who told him to read every worthwhile book twice. He entered a seminary hoping to follow his father's clerical profession, but he soon became fascinated by science after reading Darwin and the Russian Sechenov's *Reflexes of the Brain.* He attended the University of Petersburg, graduating in 1875. Pavlov took his M.D. degree in 1883, and his thesis was entitled *The Centrifugal Nerves of the Heart.* He traveled to Germany and spent two years there (1884-86) studying cardiovascular medicine and physiology.

His principal work, *Conditioned Reflexes* has the subtitle of *An Investigation of Physiological Activity of the Cerebral Cortex* (1927), which places his experiments into a relationship with the brain and nervous system, and therefore with primates and other animals.[15] Pavlov won the Nobel Prize in 1904 for his pioneering work in shaping behavior.

It is fitting that comparative physiological experiments on higher order animals, itself based on the theory of evolution, should become a part of the foundation of behavioral psychology. Pavlov was interested in the study of the two hemispheres as the controlling physiological mechanisms of behavior. He was not interested in the subjective mental states of higher order animals, though he was aware that higher order animals, including humans, had greater complexity in their reflexes.

Pavlov realized the complexity of the central nervous system and the countless stimuli impinging on the animal and effects on the total organism, and was aware that the attempt to describe cause and effect relationships given this complexity was nearly hopeless. He also realized that the presence of the experimenter in the same room with the animal had some effect on the experiment, and this is why a special laboratory for his experiments was constructed.

Pavlov knew that internal organs are regulated from the brain. However, the autonomic nervous system—that part not under direct, conscious control, such as the regulation of the pancreas, the dilation of blood vessels, control of stomach acids, or hormones—seemingly regulates itself. The operation of the autonomic nervous systems usually requires an internal imbalance in the chemical structure such as insulin in the pancreas, and not an external stimulus, to initiate a response.

Thus, some nerves do not need to travel to the brain to operate a reflex action. Touching a heated stove, for example, causes a quick motor reflex reaction that does not actually go to the brain. We usually distinguish between voluntary and involuntary or instinctive reflex actions such as a sneeze. Pavlov called these latter "inborn reflexes." However, certain reflex actions can be conditioned to be automatic, or seemingly involuntary, through conditioning, or habit formation. The association of a particular stimulus with a particular response, in order to bring about a conditioned response, is the essence of Pavlov's investigations that linked his conditioned responses to the cerebral cortex.

The classic study by Pavlov consists of the experimenter placing food in or near a dog's mouth and observing that salivation occurs. Prior to the experiment, Pavlov had performed minor surgery on the dog to divert the salivary glands near its mouth to the outside skin, as he had earlier created a small pouch for a stomach to experiment with a dog's involuntary reflex digestive tract. Pavlov then measured the salivary drops of each stimulus after repeated trials.

Next, he introduced an associated but not necessarily related stimulus, such as a light, or a tuning fork, with or without food to test whether the animal associated the new stimulus, light or sound, with the previous stimulus, food.

The *unconditioned stimulus* was the food. The *conditioned stimulus* was the light or sound. The dog salivated on hearing the introduced sound of the metronome, for example, and acquired a reflex response that normally would be reserved for the sight of the food. In other words, a *conditioned reflex*, salivation to a light introduced before food is obtained by linking a new stimulus, the light or sound, with an *unconditioned reflex*, salivation.

An *unconditioned reflex* for Pavlov is complete at birth; it is the same as an "inborn reflex." A *conditioned reflex* is the same as an "acquired reflex," the result of the establishment of new connections in the nervous system.

The innovative design of Pavlov experiments is that it combined the unconditioned reflex, the salivation of the dog, with the conditioned reflex, now known as a conditioned response. For a conditioned reflex or response to occur, an alert state in the nervous system is necessary. Once acquired, there is, Pavlov says, a new cellular connection in one or both hemispheres.

By repeatedly offering the dogs the food over an extended period of time, Pavlov was using reinforcement as a tool to condition the reflex learning. Psychology acquired Pavlov's work into the theoretical study of behaviorism. J. D. Watson and B. F. Skinner subsequently conducted experiments on behavior and reported on the results of positive and negative reinforcements measures.[16]

Pavlov also distinguished between two kinds of reflexes involved in the response: a motor reflex and a secretory reflex. He confined his experiments almost entirely to the secretory reflex, as it was the most easily measurable because of its liquid form.

Pavlov concluded that the environment was active in evoking conditioned reflexes that were either favorable or destructive to the organism. He believed that the cerebral cortex contained two major mechanisms for distinguishing stimuli—an analytical and a synthetic part to the nervous system. He deduced that animal organisms were able to discriminate between different stimuli and differing sets of stimuli to decide which were acceptable. For example, he noted that the introduction of a different pitch in a musical tone would cause his experimental dogs to bodily react to the difference. Through repetition of the conditioned reflex, however, he could discover how the animal generalized, or synthesized similar stimuli.

Pavlov experimented with the same stimuli and conditions but with individual differences in the dogs. He determined, while reflecting on the relationship between animal and human organisms, that different habits of training, education and discipline were nothing more than conditioned reflexes.

Pavlov was intellectually active in his field well into his mid-80s, working with patients in mental hospitals and investigating the effects of sleep and drugs on human conditioning. Psychiatry has used Pavlovian principles in conditioning behavior of clinical patients. But even though later in his life he spoke and wrote about neuroses of character in animals, and even hypnotic states in a dog, he rejected subjective terms like insight and understanding in describing conclusions to his experiments. He believed that subjective entities were unknown and could only be understood through quantification. In this he clashed with cognitive and Gestalt theorists for rejecting the concept of behavioral association. However, he continued to use association language to describe the "connection" between behavioral responses. He succeeded in getting the conditioned reflex adopted as the model for how learning occurs, and for the quantification that describes the relationship between stimulus and response.

He was a genius in reducing a complex set of organic reflex behavior to the simplicity of an experiment. His work ultimately instigated tremendous responses in the psychological community which tended to abandon the method of introspection and consciousness and to focus on how organisms behaved under certain conditions.

He was courageous in the face of political persecution, and critical of the Soviet authorities from the Bolshevik revolution in 1917 until 1935 after Hitler came to power in Germany. But the Bolsheviks saw the concept of a conditioned reflex as a convenient tool for reshaping society, and so luckily for Pavlov they encouraged his work and status. He died of pneumonia, February 27, 1936, at the age of 87. It was not just coincidence that the Soviets were the first to introduce a dog, Laika, in the early 1960s in the first space flights.

WOLFGANG KOHLER (1887-1967): BRING UP THE CHIMPS

Wolfgang Kohler received his PhD in 1909 at the University of Frankfurt. From 1913 to 1920 he was Director of Anthropoid Research for the Prussian Academy of Sciences at Tenerife in the Canary Islands off the coast of Africa. There he had the unique opportunity to conduct experiments with chimpanzees on problem-solving skills. His major work, *The Mentality of Apes* (1925), is the documented result of these chimpanzee experiments, a book that forced researchers to re-consider precisely what human learning was and how similar to other primate learning.[17]

Kohler returned to Europe in 1921 and became a Professor of Philosophy and Head of the Psychological Institute at the University of Berlin. He fled Nazi Germany in 1935, and for the rest of his career was a Professor of Psychology at Swarthmore College in Pennsylvania.

> *The chimpanzees manifest intelligent behavior of the general kind familiar in human beings.*[18]

Kohler's chimpanzee experiments took place in Tenerife during the period of 1913-17. He had two forms of experiments: the box problem and the stick problem. A banana was placed out of reach of the chimpanzee at the top of the cage. But a box was also placed in the cage that could be used as a jumping platform after it was moved by the chimpanzee under the banana lure. Some chimps leapt from the box but were unable to reach the banana. But Sultan, Kohler's most intelligent chimp, moved the box and placed it beneath the banana in order leap to reach it. Others chimps, observing this behavior, were thus able to turn away from the banana and towards the box as a tool which could solve their problem for the out-of-reach food.

Kohler suggested that this solution was a process called *insight*, by which after higher organisms have actively tried to solve a problem for a prolonged period of time, a sudden solution or mental discovery occurs in a time of relaxation. Kohler defined insight as: "Insight is insight into relations that emerge when certain parts of a situation are inspected."[19] This occurs sometimes after a period of reflection. He wrote: "After periods during which one has actively tried to solve a problem, but has not succeeded, the sudden right organization of the situation, and with it the solution, tend to occur at moments of extreme mental passivity."[20]

A second experiment involved two boxes in the cage, and the mental process involved was much more difficult because it involved stacking the boxes in order to reach the banana, and incorporating the second box as a part of the solution. The stick experiment required the chimps to discover using a stick to reach food out-of-reach outside the cage. Once a chimp discovered that the stick could reach the food, they sought it eagerly from one another.

Kohler then introduced two sticks, and moved the bananas further away outside the cage so that one stick alone could not be used to reach them. Sultan discovered the process of joining the two sticks, much like a fishing rod, and this discovery allowed him to reach the food. But more importantly he and the other chimps were able to repeat the discovery of the joined sticks as an extended tool over and over again.

Kohler says of the results of the stick experiment: "The stick has now acquired a certain functional or instrumental value in relation to the field of action under certain conditions, and that this value is extended to all other objects that resemble the stick, however, remotely in outline and consistency—whatever their other qualities may be."[21] In other words, one mental characteristic of these chimps is that they had learned to generalize from objects and experiences.

Kohler's *Mentality of Apes*, when first published, directly challenged the trial and error of behaviorism by proposing that some mental processes were insightful and did not need trial and error for learning to occur. Others argued that the chimps fumbling around with the boxes and sticks was indeed a kind of trial and error. Kohler tended to play down the controversy between the two theoretical camps of psychology in the 1920s by maintaining that all were looking at the same phenomena (introspection + behavior) but also at an "insolvable metaphysical problem."[22]

Kohler's observations and experiments received confirmation from the fieldwork with gorillas by Jane Goodall in Rwanda in central Africa over a period of more than two decades beginning in the 1960s. Goodall observed similar intelligent behavior among gorillas, like using sticks as tools to extract ants from an anthill, revising the idea that only *Homo habilis* was the sole primate tool-maker.

In many ways these experiments with cousin primates highlights the distinction between learning and perception. Perception is the main field of study among Gestalt psychologists. But then it was the principal field for experiment among all the first psychologists in the late 20th century. Kohler did not address the issue of development in his experiments. For example, might not older chimps do better than younger ones?

Kohler's work, like that of the first Gestalt psychologists ("The whole is greater than the sum of its parts") considered how the perception of whole systems happened physically in the nervous system, and how physiological processes in the brain were related to "structured functional wholes."

KONRAD LORENZ (1903-1989): FOLLOW ME

Konrad Lorenz was an Austrian zoologist and ethnologist who studied what he called *imprinting*, or the genetically coded behavior of a recently born mammal to bond with the first moving thing it sees as if it were its mother. He shared the Nobel Prize in physiology and medicine in 1973. He began studies in medicine at Columbia University but returned to his hometown of Vienna where he earned an MD degree in 1928. He also received a PhD degree in 1933. He was drafted into the German army in 1941 and was assigned to work as a military psychologist conducting "racial" studies, the exact nature of which is unknown. He was sent to the Russian front where he was captured and held as a prisoner of war from 1944 to 1948.

He founded the study of ethology, or the study of animal behavior. He appeared to hatched geese when they hatched, and absent their mother, the geese, thus imprinted, followed him wherever he went. His work showed how similar animal behavioral patterns imitated human primate inherited behaviors and how genetic traits govern the autonomic nervous system.

One of Lorenz's most notable books is *On Aggression* where he investigates the defense and aggressive tendencies of a variety of animal organisms, such as rats and geese, to reveal how a species fights to sustain itself and/or its species, tribe or colony.[23] Aggression and defensive mechanism are innate in all organisms, including marine life, to preserve each individual and its species from predators.

Let's look at a related human interest and pursuit to demonstrate a uniquely human primate development from Lorenz's example with animals.

The universally accepted and predominant goal of human primates is the pursuit of money. The acquisition of money automatically leads to the purchase of goods that will sustain existence and hence our survival. We are imprinted, in Lorenz's sense, with the urge to gather necessities for survival, just like all other lower organisms.

The squirrel needs to gather enough nuts to last a long winter for survival; the bear to store enough fat to keep it alive through hibernation. All nature is instinctively impelled to store enough food to help make it through the lean seasons and years. Money is the vehicle of self-preservation and not primarily a means for finding happiness or any other virtue. The more natural conclusion is that humans are acting just like other organisms in protecting themselves from future harsh conditions, whether frigid winters, floods, long droughts, or unforeseen disasters. We have been accustomed to prefer the religious perception of greed as a motive for acquiring money, and not the more obvious justification of self-preservation and species survival from future calamities.

Normally, we are born as male or female and with the individual physical and neurological distinctions that mark us for life, and we have had no choice in what those differences are. In whatever circumstances we find ourselves in, we seek to survive, to preserve our composite DNA. These innate characteristics, like reinforcing pleasurable activities through behavior, or like that of seeking money, or defending the fragile Ego from hurt, define our psychology. They are primate inheritances, the result of the long evolution of other organic developments since the beginning of all multicellular life.

Chimpanzee, National Park, Islamabad, Pakistan.

Chapter 7

THE ECONOMIC IMPERATIVE

No society can surely be flourishing and happy, of which the far greater part of the members are poor and miserable.

(Adam Smith, The Wealth of Nations, Book I: 8.35)

As human primates we have moved beyond family, tribal and clan associations—social characteristics of non-primates as well—to larger societal and national groupings. In widening our social and organizational associations into national and international settings, however, we have not diminished our primate genetic or cultural heritages and the typical social and behavioral responses. We do adapt relatively easily to technological changes that seem to improve our lives, but still might swear at a computer when it fails to function as it should. We cannot disassociate ourselves from primate origins just because we are members of large institutional groupings.

So in these final chapters I explore our economic and governmental imperatives and how they have impacted our lives, sometimes in unintended ways. For example, as individuals we have no control over large organizations, like governments or corporations that control much of our personal existence. Our behavioral responses, however, are nevertheless imprinted to react in the same way as we might to individual or collective threats, for example with fear, disgust, suspicion, animosity or hate, or helplessness, the same emotions other primates feel.

The primary twin primate economic urges I would emphasize here are the individual acquisition of goods to survive, and the fear of not having enough to survive. These dual instincts drive both individuals and corporations. The difference is that unlike individual persons, corporations do not share a common bond to help the group survive; they act only as frightened and fearful individuals do, unbound by a family or group social bond.

In this chapter I propose that the economy that influences us so directly also causes raw primate emotions like helplessness and frustration to emerge, especially when we have to speak to a respondent in India about a technical problem in our account. Social complexity has compelled us to be far removed from the source of some of our anxiety but has not lessened our primate responses. So we explore who we are as cogs in the global and national financial system. Too often the amount of money we have, or think we should have to enjoy a reasonable lifestyle, determines our behaviors, state of mind, and how others view us.

Money! There's nothing in the world so demoralizing as money.
Down go your cities, homes gone, men gone, honest hearts
corrupted, crookedness of all kinds, and all for money.
(Creon in Sophocles' Antigone)

Less than a hundred years ago America began the great transformation from an agricultural nation to a manufacturing one. It happened suddenly during World War II when the federal government funded massive projects on behalf of the war effort, building ships, planes, military vehicles, employed the jobless from the Great Depression, and ordered the nation to save on food, gasoline and necessities to support that collective endeavor. It was a time of government economic stimulation and national sacrifice, and we haven't had anything like it since.

My family saved tin foil, metals, and bacon grease for the manufacture of nitroglycerin for explosives. We planted victory gardens in our backyards to save on food costs. Everyone had monthly ration stamps for meat, gasoline and dairy products. We all collectively did our part for the war effort and were proud of it. Today it is difficult to rally that same kind of national support, though our economic situation to the lingering effects of the Great Depression is similar.

My family was self-sufficient in food, never buying meat at a grocery store, but eating what we killed as game or caught as fish. Our protein sources were venison, elk meat, pheasants, duck and geese, and salmon and trout. A refrigerated storage unit downtown held our butchered animal kills, and an icebox on the back porch kept cold perishable items like milk. The ice truck came by twice a week with blocks of ice.

We were literally hunters and gatherers in the tradition of more primitive societies, surviving by our wits and natural wisdom about the use of hunting dogs and animal behavior. Our guns and fly rods were essential tools. We picked fruit from local farmers and kept enough for the family as partial payment to can and preserve for winter consumption. We survived literally from the land without a refrigerator or central heating. I would classify us as a bib overalls social class, below blue-collar workers.

Prices have of course skyrocketed over the years, and kitchen appliances saved once time-consuming chores. America leapfrogged through the manufacturing era until the Japanese economy bested us in the 1980s, and the Chinese rose from a peasant economy after the Mao years to global economic strength in the early part of the 21st century. Our present technology-driven economy has lifted the financial boats of the more highly educated, while the majority of the population has sunk into menial, low-paying jobs, depleted what minimal savings existed, and dashed hopes for an improved future. Let me outline the figures from 2015 so the scary financial status of Americans is clear before exploring a range of economic issues.

- Average credit card debt among indebted households: $15,263
- Average credit card interest rate: 14.95%
- Average mortgage debt: $147,591
- Average outstanding student loan balance: $31,646
- Average auto loan debt: $30,738
- Only 59 percent of Americans have at least $500 in a savings account

Whether or not you are one of these average indebted Americans, you have to be concerned about the long-term financial predicament of those likely to have debt and insufficient savings at retirement. Our animal nature typically compels us to hoard, like chipmunks, the resources that will see us through the times of financial drought. But easy credit resulting in debt allows us to put off that instinct for more immediate gratifications.

Fear and economic insecurity are prevalent. Historically, the human primate feared nature the most, from being eaten by lions and tigers and bears while out hunting, and from nature's other natural disasters, like floods, fires and attacks from aggression from other tribes. With the rise of capitalism, the sources of income were jobs generated by manufacturing. The possibility of the loss of a job created its own fear and insecurity, resulting in the creation of protective agencies like unions, and the passage of laws protecting the worker.

Corporations exhibit the same instincts, often setting profits above human safety. As one example, over 50 million vehicles were recalled just in 2014—like the Ford Explorer rollovers, the Toyota unintended accelerations, and the General Motors faulty ignition switches—all defects concealed by corporate executives for years. Volkswagen software turned on its emission requirements only when emissions tests were conducted, and bypassed rules under the *Clean Air Act* when the vehicle was otherwise operating.

When concealment to protect profits that harm human safety is the priority, corporations reveal their true self-preservation and defensive motives. We prefer to call this greed. But whether we call grabbing all we can by hoarding, the acquisitive tendency is rooted in our individual nature, and by extension, into the corporate lifestyle.

Economic Decline: Bye, Bye American Pie

The world economy suffered beginning in 2008 when global economic growth became sluggish, starting in America and drifting into Europe and Asia. American economic growth was only between 1%-2% in the Organization for Economic Cooperation and Development (OECD) countries after 2008. It was the worst recession in modern history. Growth is forecasted to rise no more than 1.5% for the foreseeable future.[1] Public consumption and expenditure was in the negative range resulting in a downward trend in funding for domestic infrastructure. Less tax revenue because of the loss of employment, combined with decreased governmental support, also led to steep declines in most publicly funded activities like schooling.

Economic growth worldwide had been extraordinary prior to 2008, literally doubling from $31 trillion in 1999 to $62 million in 2010. Yet, $50 trillion was also lost in 2008-2009, an amount unlikely to be regained any time in the foreseeable future. America preached a capitalistic religion to the rest of the world, and the globalized world got the message that using increasing amounts of debt to increase assets was necessary. But the result for everyone was a delusion of gigantic proportions.

It is easy to claim that America was in a state of unabashed and unrestrained financial decadence, living on debt. But this can also be considered a part of personal liberty and a shibboleth of the radical conservative movement: to conduct one's life without restraint—to spend more than to save, and to respect the free market more than government regulations on corporate improprieties. This condition of the unfettered free market, combined with personal aggrandizement, can equally be viewed as a collective lack of public and societal discipline, to emphasize personal liberty at the expense of the common good. The result was a significant decrease in public investments. This downturn in economic strength—mostly a housing bubble fueled by unrealistic claims of permanent rising housing values and repackaged financial loans, and an unregulated financial industry—resulted in a widening chasm in income equality.

RESULTS OF INCOME INEQUALITY: HELPLESSNESS AND FEAR

As soon as riches come to be held in honor, when glory, dominion, and power followed, virtue began to lose its luster, poverty to be considered a disgrace, and blamelessness called malevolence.[2]

Prosperity usually implies wealth. But what does having wealth imply? Often, it is used to feed runaway self-gratification and intemperate desires, ostentation, and showy extravagance. Having wealth does not usually lead to investment in ideas, life improvements, or public welfare, except for rare individuals, and it does not automatically lead to an increase in jobs. Therein, partially, lies the tale of American decline. Conservative government policies and laws have lowered corporate income taxes and personal taxes for corporations and the wealthy. By default, therefore, income inequality has burdened the poor and middle classes with the majority of taxes necessary to sustain a host of public expenditures. The residual emotions are helplessness, fear, frustration and anxiety.

Nothing speaks more loudly about income inequality in the U.S. than the comparison between minimum wage laborers who work full time, and bonuses paid to Wall Street bankers. In 2014 the total bonus pool paid out to bankers was roughly double that paid to all minimum wage earners. In other words, 167,000 bankers were given $28.5 billion, or an average bonus of $172,860.

In the 1960s, chief executives were paid roughly 20 times as much as their employees, compared with nearly 300 times in 2013, according to the Economic Policy Institute in 2014.

According to the Bureau of Labor Statistics there are slightly more than one million full-time workers paid $7.25 an hour or less. I asked a waitress in 2015 how much she was paid, and she said $5 an hour, because management claimed she made enough on tips to cover the rest of her salary. Service workers tend to work around 40 hours a week on average. If they all earn $7.25 per hour and work 50 weeks per year, the total income of this group comes to nearly $15 billion, just less than half of the banker bonuses.

The richest 100 people in the world have more money than half the world's population, or the poorest 3.7 billion people. We can glory in our technological achievements in medical science, genetics, robotics, computer technology and engineering. But these accomplishments do not trickle down to everyone, but only to those individuals and countries that can afford them.

A remarkable 95% of income growth has gone to the super-wealthy. The bottom 90% of all American wage-earners—those making about $31,000 a year or less—have seen a decline of 24% in their income.[3] Touting major corporations and businesses as job creators is bogus, and the data belie the specious argument. Corporations have in fact shipped jobs overseas, hidden taxable earnings overseas, demanded lower tax rates on what they do earn, and in an arrogant display of excess given huge generous bonuses to executives and shareholders.

The premise of most radical conservative and Libertarian thinking is that government should have no business in business. Let's assume for the sake of argument that that should be the rule, because ever since the Reagan Administration's attempts at reducing government regulations, especially over the financial sector, that has been the operative policy. What have been the results? Answer: the creation of an economic aristocracy of plutocrats and corporate allies.

Here are five of the major consequences of that policy.

1. Jobs have drifted overseas, not just because of technological advances, but because corporations sought lower labor costs, no contracts with unions, and negligible benefits to workers. Between 1999 and 2008, prior to the financial collapse, more than two million jobs went overseas.
2. The tax code has been skewed primarily for the rich. By 2012, there were 116 million tax filers, of whom 58 million, exactly half, were below Middle Class incomes. In 1955 the 400 richest households paid 51.2% if income in taxes. A half century later in 2007, the same income group paid only 16.6% in federal taxes. The top tax rate in 1981 was 50%. Now it's 35%. In 1980 the maximum tax rate on capital gains was 28%. In 2012 it was 15%.
3. Special interest lobbying groups continue to write legislation that favors their products or operations, and not current benefits to workers, including the erosion of pension plans, and lack of sensible leave policies for families with children. Recall also that President George W. Bush sought to privatize social security.
4. Radical conservative groups attempt to reduce or eliminate programs that benefit workers, children, the disabled and women.
5. This self-aggrandizing, wealthy ruling class, with help from congressional advocates and sympathizers, decry the national debt, and bulging interest on that debt. Yet they helped create the debt by lobbying for and achieving reduced taxes on themselves, for outsourcing corporate jobs overseas, and hiding funds in offshore accounts. The *Jobs and Growth Tax Relief Reconciliation Act of 2003* during the Bush Administration accelerated the decrease in taxes that benefitted the rich exclusively. The result of these actions is a substantial reduction in both jobs and needed taxes in the U.S. for public services.

If the present convenient marriage between congressional and corporate policy, judging by the results, appears to be favor the wealthy and corporations, then the next question to ask is: What should be the legitimate legislative role and economic policy initiative in forging an equitable society that aids the constitutionally-required "general welfare" for working Americans and builds America's capacity to compete? The Troubled Asset Relief Program (TARP) bailed out banks. There was no corresponding Troubled Homeowner or Worker Relief Program.

Let's start to answer some questions by examining workers and wages, as worker inequalities top the list of economic issues.

Wages for Workers

There has always been a gap between the wealthy and the poor, but wages for the working class were flat for the seven years between 2008 and 2014. Salaries for white-collar workers have also been stagnant, stuck at $455 a week (in 2014), equaling $23,660 annually, below the poverty line for a family of four. Workers are not only not highly paid, they do not control their own time on the job to receive overtime pay. Corporations slashed worker pay and benefits, and eliminated competition through price wars and mergers. It also allowed Wall Street firms to experiment excessively with new kinds of securities that were speculative but bordered on gambling. We discovered what these financial devices were not in 2008—they were not collateralized debt obligations, securitized mortgage-backed assets, or derivatives.

Even Alexis De Tocqueville in 1838 had a traditional insight in the necessity for the slow growth of wages:

> *I think that, upon the whole, it may be asserted that a slow and gradual rise of wages is one of the general laws of democratic communities. In proportion as social conditions become more equal, wages rise; and as wages are higher, social conditions become more equal.* (*Democracy in America*, 1956, p. 227)

Nothing seems to make for more sense economically than that all workers receive a just wage, and that income inequality not be allowed to dominate the whole economic landscape.

De Tocqueville was not alone in his analysis of how taxes should be laid or how they wages should be distributed. Alexander Hamilton in Number 36 of The Federalist Papers pointed out that taxing the wealthy was judicious for the national good.

> *The quantity of taxes to be paid by the community must be the same . . . and must naturally tend to make it a fixed point of policy in the national administration to go so far as may be practicable in making the luxury of the rich tributary to the public treasury in order to diminish the necessity of those impositions which might create dissatisfaction in the poorer and most numerous classes of society.*
>
> (*The Federalist Papers*, p. 222)

Hamilton did not have the fluidness or clarity of Madison in his writings. But he clearly saw that the rich would have to pay their fair share, and pointed to the possibility that an inequality of taxation would upset the lower classes, and that that would not be a welcome prospect.

The present income gap has widened beyond normal expectations and threatened the livelihoods of the majority of workers. As of 2015, the average family income was $52,300, or about where it was in 1996, and a decline from its high of $56,000 in 2007. There is also a sizeable gap between ethnic and racial incomes, a situation that generates its own imbalanced social dynamic. According to the Federal Reserve, incomes dropped between 2010 and 2013 from 2-7% for all income groups except the top 10%.

A *USA/Today* poll in 2014 found that the true cost for a modest living standard for a family of four comes to $130,357. Yet only 12% of American households had achieved that annual income stream. According to the Gallup Poll in 2014, Americans reported spending more than they did in 2013, but this increased spending was on household essentials like groceries and gasoline instead of discretionary items such as dining out and travel.

The severe decline in wage-earners incomes now constitutes about 50% of all workers in America, an economically unhealthy status that in past eras resulted in revolts, certainly the Peasant Revolt in England in 1381 and the French Revolution in 1789. *The top 10% of Americans control half of all income generated.* To show how this imbalance has changed so dramatically, the majority of workers in 1960 controlled 65% of all income. The top 10% of wage-earners had only about 20% of all income. This dramatic reversal of income results in loss of government revenue for the sustenance of vital public programs. History has consistently demonstrated that internal changes in a society can shake the social foundations often more profoundly than external aggressions.

A traditional view is that wealth inequality will eventually stabilize over the years and that no state intervention is needed for this to occur. Historical analyses, however, reveal that the concentration of wealth does not automatically equalize. This conclusion is true also in other developed countries. Universally, income from wealth grows faster than wages. The concentration of wealth also implies that the wealthy will have more influence on the political system and manage elected officials to favor its class.

A *Survey of Consumer Finances* by the Federal Reserve Board in 2014 indicated that that a median family headed by a householder under 35 years of age earned $35,509. That amount is 6% less than earned by similar families in the first survey in 1989. Younger families are declining in income further and further behind older families.[4]

Capitalism and Inequality

If you are poor, you'll always be that way;
Only the rich get richer every day.
(Martial, Roman poet 40–104 CE)

The three dominant political and economic issues of contemporary western societies are 1) the rapid growth of economic inequality, 2) the corresponding stratification of society, and 3) the role of government in correcting or modifying any imbalances. The conservative political right wants to curtail spending and taxes to keep the economic machine running. The more liberal left seeks to increase government spending to stimulate slow growth, and to help benefit those on the lower economic strata. But both sides of the political spectrum are not considering the rising tide of social instability that can generate a populist backlash creating social disorder and possible chaos.

It was only the creation of the modern welfare state in America, chiefly during the Franklin Roosevelt Administration that established educational, work, and social programs, that allowed democracy and capitalism to co-exist in a relative harmonious state. It is reasonably argued that legislated rewards—like unemployment insurance, food stamps, and Medicaid—to the economically disadvantaged creates injustice among those who believe that the so-called government handouts are not earned through merit.

According to the Census Bureau, 10% of all households were on food stamps in 2004. By 2013, that percentage had doubled to 20% of all households, primarily because of wage inequalities. The cost of the government program increased 42% from 2009 when the recession hit with full force to 2013. This is why tax-cutting conservatives sought to cut the program's beneficiaries as a curtailment of big government, a slogan used to slash all social budgets, but not the military and defense. One quarter of all children benefit from food stamps, an awesome burden for the families.

On the other hand, maintaining a dynamic economy though capitalism is only one leg of a modern democracy. Providing benefits as a percentage of the economy's total national strength is the other leg of the economy, the price a nation has to pay for aiding the lives of citizens who fear increased disenfranchisement and insecurity. Conservative Congresses have systematically adhered to ideological priorities like: tax cuts for the wealthy that presumably help the economy and trickle down to workers, that austerity floats all boats, and that the markets regulate themselves without government interference.

Schemes based on these premises have been discredited. The other discredited notion is that somehow presidents unduly influence economic outcomes.

Thomas Piketty's 2014 book, *Capitalism in the Twenty-First Century*, which reveals economic data over the last 300 years, shows the negative influence of high rates of income inequality. Nobel Prize winner in economics Joseph Stiglitz, made a similar case in his book *The Price of Inequality*.[5] Fragility in economic growth, widening income disparities that create an aristocracy of wealth, and reluctant political will to change policies has resulted in apathy about slowing the measured income inequality.

> *The passion for wealth takes the place of ambition, and the heat of faction is mitigated by the consciousness of prosperity.*[6]

The six heirs of the house of Walmart together hold wealth of $69.7 billion, equivalent to the collected wealth of the bottom 30% of the U.S. population. In January 2015 Walmart terminated the health coverage of 30,000 part-time workers, about 5% of the total workers, who worked less than 30 hours a week. Walmart employs 1.3 million in the U.S. Larger numbers than expected enrolled in health care plans after the passage of the *Affordable Care Act*, so businesses like Walmart, Home Depot, Target and Trader Joe's found their costs rising.[7] Corporations are about making money for shareholders and themselves, and workers are necessary expenditures whose health care can be passed on to the public, and whose wages can be minimized when necessary.

According to the National Research Council, America has lagged far behind other industrialized countries in health indicators such as infant mortality and life expectancy. These factors can be traced to income inequality and the rise of poverty. Health problems of the poor fall heavily on the health care industry, one reason why costs are rising. The poor are largely non-white, unmarried and poor single women with little prospect of moving up any economic ladder. Some of the blame falls on the economy itself. But a large part is due to radical legislative intransigence toward this segment of the population and its fixation with cutting taxes and reducing limited government payments to the poorest citizens.

All rights are conditional and derivative from the social compact, however determined, and not just the legal rights in the *U.S. Constitution*. Neither government rights nor individual rights are absolute. Rousseau's *Social Contract* (1762) describes this social obligation between citizens, a book that helped fuel the French Revolution and bring down royal sovereignty and the divine right of kings.

America's economy had crystalized at the beginning of the 20th century based on capitalists of enormous financial appetites. Our economy, then as now, maintains the same elements—the ownership of production, the exploitation of labor and of natural resources, whether in oil, coal, timber or food products, and large-scale distribution that created the first mass consumer culture.

This is still our legacy, together with the vast sums spent on lobbying politicians for industrial interests. That legacy is built on the slogan of *laissez-faire* capital economics first proposed by the robber barons as a way of protecting their money. It is not based on the constitutional pledge to promote the general welfare, which presumes that the public's good is preferable to private gain. The era of the industrial tycoons, from the end of the Civil War to the beginning of the 20th century, had the greatest level of income equality with ostentatious prosperity and glaring poverty. America is now approaching that level again.

Those with even a smattering of history will realize how our tradition of extreme wealth in the hands of corporate executives, just like it once was with the robber barons, is easily discernable. It is not voodoo logic, something conjured as a puzzle like that proposed by the witches in *Macbeth*. Our economic system bears all the hallmarks of wide divides in the distribution and ownership of wealth. Corporations claiming that their first claim is to shareholders, whose motive also is only financial gain, is reasonable but unconvincing as service to the public.

It is an illusion to think that income inequality will become more equitable, or that the laws of the marketplace, besides the profit incentive, will somehow decrease wealth inequalities and result in financial harmonies among the rich and ordinary workers. Wealth has always been in safes of the rich. It's just that in this part of the 21st century it is more concentrated than ever in the top 10% of the wealthy, resulting in more individuals and families in poverty. The so-called Middle Class has less purchasing power. At a time when consumer spending drives the economy, ordinary consumers have less funds to purchase anything.

An extremely conservative body of the U.S. Congress has prevailed in preventing any even modest increase in taxes for the wealthy, and hampered efforts to regulate the financial industry that appears to be strangling ordinary laborers. Worse, they declare publicly that the unemployed are somehow lazy and unwilling to work, yet never cite any research study or evidence to bolster that claim. Disparaging the poor and unemployed is a convenient way of getting like-minded conservatives to vote against government programs that aid the economically less fortunate whose condition is by and large not their fault but that of the depressed economy.

The decline in education is the most distressing. Millions of Americans are poorly educated working class men and women. About 25% quit school before high school graduation. Less than high school graduates have only a 20% chance of earning a minimum wage. They cannot find good jobs. The racial and social class divide is apparent to anyone who follows the demographic numbers.

The economic decline impacts education funding at all levels, and is partially responsible for lowered educational standards and negligible test score results. Just as wealth has increased for the wealthy and decreased for the poor, education attainment levels between the wealthy and the poor have also widened. The gap in the United States is about 40% larger than it was in 1980.

The average spread on a standardized test at this writing is about 125 points between the top 10% of children from rich families and the bottom 10%. This point spread is more than twice the gap between White and African American students. Moreover, the divide is not just between the rich and the poor, but also between the rich and the middle classes. Family income is the chief predictor of schooling success. Income and educational inequality is a global problem. But its damaged image in the U.S. has enlivened educational and political debate about the loss of American pre-eminence.

According to the OECD in 2013, one in six 15-29 year-olds was not in employment, education or training in the United States, a higher probability than on average in OECD countries. Conservative governments are thus more likely to have lower tax revenues, have higher welfare unemployment and payments, and greater social instability because of unemployed and demoralized youth. The other part of the equation is that most youth do not have the skills essential for a global labor market.

JOB LOSS AND ECONOMIC DECLINE

It is the loss of jobs that speaks volumes about where America is heading. The loss of manufacturing jobs to foreign countries is over simplified. Technology, not just cheaper labor, has changed how industries compete everywhere. Businesses have adapted to the technology, like the use of robots, and can be just as efficient with fewer employees. Meanwhile, more and more workers have skills that are outdated for the more sophisticated technological world.

According to *Time* columnist Rana Foroohar, 58% of the jobs created since 2010 have been in low wage positions, mostly in food services and retail stores. The salaries of these workers are less than $13.83 dollars an hour.

Highly educated and highly skilled workers, the winners in the schooling arena, will move to where the higher-paying jobs are, and not just in the U.S. Most workers cannot or will not relocate, which means lower- and middle-class Americans will suffer the consequences of globalization and the slow diminishment of wages. But capital moves at electronic speeds and successful businesses adapt to global markets with technological efficiency and with fewer workers.[8]

Fareed Zakaria claimed in The *Post-American World: Release 2.0* that the post-American world is not about America in decline but the rise of other countries, what he called "the rise of the rest." According to the *United Nations 2013 Human Development Report*,[9] China has surpassed Japan as the world's second largest economy. India, as the world's largest democracy, is reshaping its economy and social policies. Brazil has significantly raised its living standards. Indonesia, Turkey, Mexico, South Africa, and Thailand are becoming more economically creative and inter-dependent.

Zakaria claimed initially that the new world order did not herald that America was in decline. He modified that view where he acknowledged that America was in "relative decline."[10] America is declining slightly economically, as are Europe and Japan, two of the world's strongest economies. In early 2014, Europe had a 12% jobless rate while America's was slightly less than 7%. Europe will have a longer and more painful recovery if it can sustain the value of the euro currency, especially in the Mediterranean countries. America has a slightly higher growth rate than Europe and a much larger one than Japan. In these times of mediocre economic recovery, that does not argue for enhanced optimism.

About 60% of the newly created jobs go to hourly wage-earners receiving, if they are fortunate, about $15 an hour, such as those with good jobs in car manufacturing plants. Raises in the minimum wage will hardly cause a ripple in the economy, but will give a breath of relief to those in the lower income earning class.

Here's a comparative difference compared to the economy in my youth. In 1950 as an occasional fruit picker and part-time day laborer, I earned $.75 cents an hour, then the standard hourly wage. That doesn't sound like much, but the average annual salary in 1950 was $3,000. If I had worked fulltime, earning $1,440 for a year, I would have earned about half of the average annual salary.

Today, in 2015, the hourly wage is pegged at $7.25 an hour. A fulltime employee at that hourly wage would earn $13,920 annual salary, far below a poverty level wage. But today's hourly wage annual salary is four times lower, or only about 26%, of the average $52,300 annual income. In sum, compared with more than 70 years ago, wage earners earning only an hourly wage today are twice as poor comparatively as were their grandparents like me in their time.

The top 5% of all wage-earners accounted for nearly 40% of personal consumption expenditures in 2012. The comparative ratio was only 27% in 1992. The top income earners are driving the economy, not the middle or lower classes. The top 20% of all households generated about 90% of all increases in inflation-adjusted consumption between the recession years of 2009 to 2012.

And consumers rely chiefly on advertising, which is omnipresent and persistent. Advertisers use standard image relationships and behavioral modification techniques to sell products. Few would automatically equate the idea of love with a car. But if the reinforced association—or the conditioned reflex of Pavlov—is seen often enough the image becomes more or less habituated and strengthened in response. Sexually provocative images sell beer and food items, just as happy people with dogs, children and the outdoors sell drugs. Associating images is a profound primate capability, and not just for humans.

Moreover, TV is the major vehicle for the promotion of commodities. Most athletic events are timed for about 60 minutes, like a typical professional football game with a season lasting five months. Actual football playing time might equal about 11 minutes, because the clock is running even when the ball is not in play. An actual play lasts 4-5 seconds. More time is shown replaying expired plays, about 17 minutes, than is actually shown in the real plays. But the TV time for the game averages 3 hours and 12 minutes, during which time there are over 100 commercials shown in about 20 commercial breaks. Commercials typically consume about one-third of the total broadcast for all television shows, during which corporations try to convince you with slick videos that you really need their product, though in many instances you do not. Repetition is positive reinforcement for a product, a strong inducement to buy.

America's primary religion, based on attendance and viewership, is professional football. As one of America's largest corporations, the National Football League (NFL) has its detractors ranging from health care professionals who complain about traumatic brain injuries in about a third of the players, to women and citizens who claim that players abuse their wives and girlfriends in domestic violence disputes. But this doesn't impede the money flow. The NFL at this writing takes in about $10 billion in revenue annually, and its 32 league owners pay the NFL Commissioner $44.2 million in annual salary. In an eggshell, that's America's values. The chief advocate of our most violent sport after boxing is worth 110 times more to us in salary than the president of the United States.

Americans are consumed with watching mostly advertising for trucks, cars, beer, food like pizza, and mobile cell phones, among other sponsored, and often very expensive, items like watches. Correspondingly, there is little widespread public dissemination of public safety concerns, like why eating too much, and too many sugar-laden and calorie-rich fast foods, might be injurious to long-term health. Corporations build in the cost of advertising into the product, so consumers, when they buy the product, actually are paying for that slice of corporate spending, and not just for the actual making of the product. Corporate values do not always favor the health of the consumer.

Where Did the Money Go?

Less than fifty years ago America made another radical transformation from the manufacture of goods and commodities to erecting newly-designed but flawed financial institutions that, purportedly, were supposed to drive economic development. Instead, skirting, flouting or reducing government regulations, the emerging banking and financial sector created vast wealth reservoirs for managers and speculators, bankrupted financial enterprises—as the Savings and Loan scandals did during the Reagan Administration—created the toxic technological bubble in the 1990s, and fueled the home loan disaster during the Bush Administration.[11]

Managing a national economy is not easy. But yielding unprecedented amnesty to banks, as the government has done for the past few decades in collusion and cooperation with financial institutions, has been a catastrophe long in the making. But legislators and bankers, and their bonding to support each other's interests, have not been the only guilty party. Consumers bought into the easy credit, into borrowed money and convenient access to debt. Student debt from college loans, outstripping in volume auto loans and credit card debt, is an onerous burden on the individual and a looming financial crisis for the economy.

The creation of credit cards, a carnival of debt extravagance, and the government's collusion in deregulating credit card interest rates, may stand as one of the most creative means by financial institutions to convince consumers that they could enjoy commodities they might not otherwise purchase. Banks got cheap loans from lower government rates, then charged consumers exponentially higher rates on consumer credit card use.

And the scheme worked spectacularly, better than a three-card monte hustle, or the house percentages on gambling in Las Vegas, because people owed banks more than just the loaned funds. They became indebted more or less permanently with higher and higher rates of interest. The nation's consumers are locked in hock to banks, unwilling to pay off the loaned money each month when it comes due.

Instead of making useful commodities people need, America's economy was built on money chasing money, or, in the obtuse terms of the malfeasance of banks, on monetary creations like derivatives, securitized loans, credit default swaps, and collateralized debt obligations. Survival, once based as mine was in youth on obtaining sufficient food sources, then later to purchasing manufactured commodities like a newer car or a refrigerator, has morphed now to surviving by getting enough funds to eliminate debt. America is now living in its own horror movie, slaves to financial interests as wages decline, businesses falter or fail, and home ownership, once the pillar of financial security, struggles.

Presidential executive orders now require contractors receiving federal funds to pay higher minimum wages, ban retaliation against workers who discuss their compensation, provide employees more information about their pay, and protect gay and transgender workers from discrimination.

Are We There Yet?

From 2008 to 2015, the GDP has risen only about half as much as the typically recovery since the end of World War II. The federal budget deficit has shrunk in half. Housing prices still remain 5% lower than in 2009. Household debt remained unchanged. According to the *Annual Report to Congress 2013* of the Federal Housing Finance Agency, the home price index remains 15% below the peak reached in the first quarter of 2007, and 5% lower than when the recovery began.[12] Dinah Walker, an analyst with geoeconomics at the Council on Foreign Relations, has drawn historical parallels with the American economy. She showed that the current recovery was far below the average of past economic recoveries and once projected recoveries.[13]

To update, the average household in 2013 had a $120,000 home mortgage and each household had an average of 13 credit cards.[14] According to the 2014 report of the Stanford Center on Poverty and Inequality, the annual median household income in the U.S. in 2014 was $52,300—as noted above—a 7% decline from the 2007 figure of $56,100.[15]

The Wall Street Journal/NBC News poll found that 76% of adult Americans lack confidence that their children will have a better life than they have had. They place blame on Washington officials. The President's approval poll number dropped, but that of Congress fell to 9%. Those polled felt that income inequality contributed to their anxiety. They viewed the state of the economy dismally and America's role in the world, and believed the political system dysfunctional. [16] The state of the economy created widespread anxiety and fear, raw emotions that trigger primate sensitivities. According to *Time* columnist Rana Foroohar, the grip on the financial sector is as tight on the U.S. economy as it has ever been. [17]

John Stuart Mill in 19th century England believed increased wages and unionization were positive actions. "If it were possible for the working classes, by combining among themselves, to raise or keep up the general rate of wages, it needs hardly to be said that this would be a thing not to be punished, but to be welcomed and rejoiced at." [18]

I cite Mill and others frequently because the great minds of the past continue to emphasize the economic wisdom that is absent in American economic policies, and it is important to question whether that is because of ignorance or obstinance to sound fiscal policy.

Mill's argument, still relevant today, is that of course the consumer must pay for the increased labor costs, and that production costs are kept low by less labor—usually because of increased use of technology like robots—but also that more workers stay employed, which is healthy for the economy and the worker's family. The issue of wages, as with similar economic topics, is whether one overly favors higher profits at the expense of humane and civil responsibilities that affect the workers. Clearly, a balance between reasonable profits for the owners of production, and those who produce the product is the most sensible legislative strategy.

In 1848 when John Stuart Mill published his seminal study, *Principles of Political Economy*, excessive wealth had its critics. He wrote: "That the diffusion of wealth, and not its concentration, is desirable, and that the more wholesome state of society is not that in which immense fortunes are possessed by a few and coveted by all, but that in which the greatest possible numbers possess and are contented with a moderate competency, which all may hope to acquire."[19]

Future financial challenges are daunting and seemingly intractable in a stymied nation with a strong anti-government movement. Oppositional legislatures are more interested in ideological purity than in national achievements. Yet According to the American Society of Civil Engineers, the condition of the nation's infrastructure receives a D grade for lack of upgrades.[20]

Goldin and Katz analyzed a comparison between high school and college-educated workers between 1890 and 2005—just as the number of college graduates began to decline—and the rate of growth of college degrees increased. They found that wage inequality in the U.S. is a direct result of the failure to invest in higher education.[21] Their research also showed that American working class citizens no longer surpass the educational attainment of their parents. An age gap between generations has emerged as well as an income gap.

One conclusion is that the best way to increase average productivity of the labor force and boost the economy is to invest in education. Instead, severe budget cuts have been made throughout all educational levels reducing the opportunities for the next generation and beyond. This setback comes as increased levels of schooling are required for more sophisticated technical positions. As taxes have been reduced, colleges, whose budgets have been severely curtailed, have attempted to maintain quality by raising tuition rates, thus reducing the numbers of potential students who could attend by pushing them out of the ranks of affordable prospects.

Percentage Share of Top 10% in Total Income

	U.S.	Britain	Germany	France	Sweden
1960	32%	29%	32%	37%	30%
1970	31%	28%	32%	32%	28%
1980	38%	32%	32%	32%	23%
1990	42%	38%	33%	32%	25%
2000	47%	40%	37%	33%	27%
2010	48%	42%	37%	33%	27%

Source: Piketty, p. 323.

Piketty's research shows that between 1987 and 2012 the number of billionaires rose from 140 to 1,400, and their total wealth grew from $300 billion to $5,400 billion. "The decrease in the top marginal income tax rate led to an explosion of very high incomes, which then increased the political influence of the beneficiaries of the change in the tax laws, who had an interest in keeping top tax rates low or even decreasing them further and who could use their windfall to finance political parties, pressure groups, and think tanks."[22]

The top one thousandth of the world's wealthiest owns nearly 20% of total global wealth. The top centile, or "five-percenters," owns about 50%. The top decile owns between 80% to 90% of the world's wealth. The bottom half of all global wealth owns less than 10%.[23] These percentages are approximations, but they serve to focus our collective attention on the global distribution of wealth.

Let me cite Mill again as a conclusion because his statements continue to be as relevant today as when composed.

> *I know not why it should be a matter of congratulation that persons who are already richer than any one needs to be, should have doubled their means of consuming things which give little or no pleasure except as representative of wealth.*[24]

The Indigent: Hunger and Other Disorders

Gannets, as they scan the ocean from their flying lofts, hungrily search for the sardine swarms about sixty feet down in the ocean, aided by accommodating dolphins that swim under the sardines to bring them to the surface. Whales too join in the feast and scoop gobs of sardines into their mighty maws.

Cranes, storks and swallows migrate over 6,000 miles from the tip of South Africa to northern Europe to return to their nesting sites, all along the route searching daily for their next meal, or to scavenge the leftovers of the kills of other predators. Crocodiles, knowing the habits of their prey, wait for the wildebeest hordes, about a million strong, to cross the Mara River during their migratory journey.

The natural world of non-human and most primate humans is on a daily and perennial hunt to find its next food source. But the human primate, disassociated from the daily search for provisions, has arranged for its food to be bought in grocery stores and restaurants. For a large number of people, because of the length of the economic recession, hunger has emerged as a natural condition that exists among other non-primate organisms in their natural habitats.

As Congress slashed programs benefiting the poor, and increasingly for the declining Middle Class too, more Americans descended into the lists of the indigent.[25] The number of hungry Americans, what the government calls the "food insecure," grew to 48 million in 2012. There are nearly 18 million households, equivalent to 16% of the U.S. population and most of them children, who have inadequate resources to meet basic food needs. Congress in 2013 slashed funding for food stamps but increased the tax benefit to farmers. Not all food grown in the U.S. is for American consumers. Much of it is shipped overseas for export earnings.

In 2014 there were 50,000 emergency food programs in the U.S. There were only a few hundred such emergency centers in the early 1980s. Eating just packaged foods and fast foods increases the risk of obesity, which goes hand in glove with poor nutrition. About 72% of all food stamp participants are children. Mississippi has the highest rate of food stamp participants at 22.5%. There are 17.6 million households, equivalent to 16% of the U.S. population, that don't have adequate resources to meet basic food needs.

In my coffee discussions with conservative, retired business executives, the federal food stamp program and similar so-called welfare programs always seem to arise. Conservatives unanimously seek to destroy or curtail it, and hate what they consider to be a slide into a welfare state. The Food Stamp program (Supplemental Nutrition Assistance Program, or SNAP) is the largest nutrition program the government funds, though farm bills are typically the most expensive domestic programs.

During the Reagan years, when America was relatively economically buoyant, the number of the very poor and financially struggling increased. One 1989 survey found that requests for emergency food assistance increased 96%. For many of the persistently poor, food stamps were literally lifesavers. Recipients have to have a monthly income 130% below the poverty line. On November 1, 2013, the Republican House cut the food stamp program as a cost-cutting measure. Those helped (47 million in 2013) were primarily children, but also seniors and the disabled.

Who are we as consumers, when too many of us are hungry? We are a broken people too, not unlike those unemployed men who came to the back door of our house during the

Depression seeking something to eat. My grandmother always gave them a sandwich and a glass of milk, but made them cut some wood and do some chore for it, and they were pleased to do so.

In other times and other places, revolts and revolutions have been the result of income inequality. Not all identical conditions apply, but the Peasants' Revolt of 1381 certainly had many painful and similar social conditions. Besides the fatal results of the Black Death, there was war, oppression, government incompetence, a weak acceptance of the system, and a profound mistrust of government. The Peasants' Revolt nearly succeeded in completely toppling the kingdom.[26]

The unequal wealth and service conditions are similar to the robber barons of the 19th century in America—Jay Gould in railroads, J. D. Rockefeller in oil, J. P. Morgan in banking, Philip Armour, Gustavus Swift and Nelson Morris in pork and cattle, Andrew Carnegie in steel, Henry C. Frick in coke for smelting iron into steel, among others. Together they controlled much of the nation's resources and industrial manufacturing.

These masters of American industry, the imperialists of their day, scornful of the laboring classes, once they had completed the conquest of the nation's resources, were pleased to declare for the sanctity of property rights that government was obliged to protect. Charles Morris in his book *The Tycoons* called their activities "bandit capitalism." These banking bandits collared most of the wealth of the world, and their unlawful marriage with political leaders debased the "inalienable rights of man," and by default denied society's less fortunate a fair share of urbanity and just comforts.[27]

It's been a century since major worker strikes against industrialists, like the Great Railroad Strike (1877), The Pullman Strike (1894), the Chicago Stockyards Strike (1904), the Triangle Shirtwaist fire (1911) that killed 146, mostly immigrant workers, when they couldn't open locked doors, and the Ludlow massacre (1914) when strikebreakers and the national guard in Colorado machine-gunned striking miner camps and set the camp ablaze killing 13 women and children.

It's difficult to predict when social conditions will reach a peak when massive demonstrations and riots that rip apart the social order. But I witnessed such urban violence in 1968 when I first worked in Washington DC. Then I saw marines in full battle gear armed on the rear of jeeps with machine guns patrolling the streets of Washington DC, as buildings burned about ten blocks from the White House.

Riots followed the assassinations of Martin Luther King and Robert Kennedy in 1968. But it shouldn't surprise anyone if some spark ignites the masses to display their collective displeasure over the erosion of jobs and spending power, and a lethal lack of governmental action to salve their economic misery. The economic pot is boiling and the American primates are angry. Their salve might not just be more promises from elected officials or candidates for offices.

Eating Away at Social Services

The conservative Congress has eaten away at the government's provision of food stamps. It has also eaten away at other social services benefitting the public.

Just short of half of all states, 23, refused to accept Medicaid payments in their state for the *Affordable Care Act*. Former indicted Texas Governor Rick Perry canceled the Medicaid program after the U.S. Supreme Court decided that states could opt out, thereby deciding not to receive billions in federal assistance. As a result, two million citizens in Texas became ineligible to receive any health benefits. In addition in 2014 Texas had six million uninsured low-income citizens, the most of any state.

After more than 50 votes in Congress to eliminate or weaken the *Affordable Care Act* (ACA) that provides reasonable health care for all Americans, and a U.S. Supreme Court challenge that sought to weaken it, on June 25, 2015 a simple 5-4 majority ruling made the law permanent, ending futile attempts to deny its legislative existence.

The lame and implausible argument against the ACA law by critics was that it was akin to socialism, a word that in the minds of uninformed Americans is tantamount to the most radical form of communism, but sadly not like those dictates in the New Testament. Socialism is associated with Karl Marx and Vladimir Lenin and the tyranny and brutality of the Soviet Union in which there was no private property, in which the state owned all means of production, and advocated class warfare on capitalistic societies. On the other hand, social programs that provide services to the public are standard throughout the developed world, certainly in Europe, in which the government owns or controls select enterprises.

Yet who initiated the Individual Mandate for health care that Republicans opposed during the Obama Administration? Answer: The conservative Heritage Foundation and Republican Senators Orrin Hatch and Robert Bennett of Utah, and Charles Grassley and Christopher Bond of Missouri in the *Consumer Choice Health Security Act* (1993).

American style socialism exists in laws passed by Congress to aid citizens to fund so-called welfare programs, like the Tennessee Valley Authority that brings electricity to rural homes, or state programs like the Alaska Permanent Fund that collects revenue from oil and gives the proceeds as dividends annually to all Alaskans.

Reasonable people (and there may be few of those left) can disagree about the degree to which Congress alleviates somewhat the misery of poor Americans, especially vulnerable children, and debate whether this is charity or simply an extension of federal aid to the constitutional mandate of promoting "the general welfare."

A corollary of economic distress is the search for relief that often turns to social disorder. A lack of education funding, and a lackluster schooling experience, can also contribute to social unrest. These conditions mean there is more use of drugs to cure pain, and more crime to obtain some essentials.

But then admittedly, there is a whole ramification of social influences. Youthful mistakes by the less fortunate turned too many into more than just excessive smokers and drinkers, producing a strain on the health care system. Many medicate with crack cocaine and heroin. More than 660,000 Americans used heroin in 2012, and the numbers have soared in the past few years. Heroin users rose 200% in Louisiana, 300% in Maine, and 60% in Ohio.[28]

The topmost marginal tax rate of the income tax in the U.S. fell from 70% in 1980 to 28% by 1988 during the Reagan Administration. Similarly, the top marginal tax rate of the inheritance tax fell from 70% in 1980 to 33% by 2013. Both of these declines in tax revenues gutted many federal social programs, but not the military budgets. What is clear, according to Piketty, is that the reduction of these and similar tax rates, and the corresponding rise in the top personal incomes, did not stimulate productivity, contrary to conservative belief that such compensations to the wealthy are equal to a trickle-down system of income and jobs.[30]

Public Wealth in the U.S. as a Perentage of National Income		
	Assets	Debt
1920	75%	50%
1930	75%	40%
1950	110%	100%
1970	125%	50%
1990	110%	80%
2010	125%	100%

Source: Piketty, p. 153

Regulating Corporations

Extremes of wealth appear to be embedded in the capitalist system, or certainly have been since the end of World War II in 1945. Joseph Stiglitz argues that it is not the laws of economics that are contributing to inequality but political policies.[31] Economic inequality has furthered political inequality as the political parties cannot reach agreement on legislative remedies.

The biggest and erroneous policy of all was the move to limit regulation of financial institutions beginning in the Reagan Administration, that witnessed the savings and loan downfall, and which eventually helped create the financial crisis of 2008. Shortly after the collapse of the economy, the banks accepted taxpayer bailout funds while the homeowners who saw their residences lose value were shortchanged.

One reasonable response to the financial crisis, and its unpopular subsequent bailouts to banks, might have been legislation that reduced the size of financial institutions based on their capital holdings. This might have reduced speculative investment tendencies. This would have meant that banks needed to hold enough cash to keep from going bankrupt, an accounting that the *Glass-Steagall Act*—enacted in 1933 to separate commercial banking from investment banking, and foolishly repealed in 1999—once required. The repeal of *Glass-Steagall* effectively created the same financial conditions that existed prior to the 1929 crash that generated the Great Depression. The financial sector earns 30% of all corporate profits but creates only 6% of all jobs.

The regulation of America's financial sector is overseen by a jumble of federal agencies: the Treasury Department, the Federal Reserve, the Securities and Exchange Commission, the Commodity Futures Trading Commission, the National Credit Union Administration, the Federal Deposit Insurance Corporation, the Federal Housing Finance Agency. Most of these agencies have overlapping jurisdictional authority, so that lobbying groups can find favorable legislation that protects their interests in specific instances.

Corporations, the largest of the special interest groups, according to the U.S. Supreme Court, are persons deserving of the same rights as biological persons. They are like those chemical corporations that lied and gave false information about their chemical dispersals like DDT into the environment, the same ones Rachel Carson exposed in her 1962 book, *The Silent Spring*.

Corporations as persons like Enron that acted as an intermediary in the energy sector that hid billions in debt through fraudulent accounting practices, became the largest bankruptcy in U.S. history, and lost the pensions of former employees.

Corporations like oil companies with lax safety equipment that spill their oil on the environment and then try to claim protection from lawsuits.

Corporations like financial enterprises too big to fail that ruin the economy, dissolve the savings of working people who trusted them with losses, then seek government funds to regain their status as protectors of the economy.

Corporations like car manufacturers that have ignition switches that switch off while driving, then fail to recall cars after hiding details from federal investigators, then, when exposed, recall 2.6 million cars to protect their image of safety; defective air bags from a supplier company in Japan that spew shrapnel, whose corporate executives in turn covered up safety failures.

Corporations like telecommunication companies that fraudulently overbill its customers for services never ordered like "flirting tips, horoscope information and celebrity gossip," a percentage of the surcharge of which goes to the phone company and not the third party.

Now corporations have religious freedoms so they can exercise their religious beliefs irrespective of the claims of their workers to not be discriminated against in health care or other government entitlements.[32]

Government is the solution to many of society's problems, despite President Reagan's claim that government is the problem. But the excesses of corporations are the problem—acid rain, chemically polluted streams and rivers, lack of safety standards in mines and oil rigs, faulty devices in vehicles, speculative financial schemes with taxpayer money, to name the most odious. Government is the only counter-weight to capitalism and to curbing those excesses, and the protector of the safety of all Americans.

It gets worse. Corporations are opting out of their civic responsibility to pay their fair share of corporate taxes by leaving the country. This is a process known as *inversion* in which businesses acquire foreign companies and then move their operations to those headquarters overseas. The construction company McDermott International pioneered the move in 1983 when it moved its operations to Panama avoiding $200 million in taxes. Nearly all telecommunication operators answering phones are already in India or other English speaking countries. More than a hundred companies have effectively renounced their business citizenship since 1983.

The curtailment of corporations seeking to create monopolies and control public access to services began in 1904 when the U.S. Supreme Court in *Northern Securities v. United States* against the financial railroad giant James Jerome Hill and the banking baron, J. P. Morgan. [33] Hill created Northern Securities as a vehicle for controlling all railroads serving the Midwest. President Theodore Roosevelt ordered the Justice Department to build a case against this monopolistic practice using the *Sherman Anti-Trust Act* (1890) as a legal basis. The court ruling was founded on the constitutional principle that the government has a right to regulate interstate commerce.

The first Roosevelt Republican Administration then convinced Congress to pass the *Hepburn Act* (1906) that granted the Interstate Commerce Commission (ICC) the power to regulate the rates, "fair, just and reasonable," that railroads could charge customers. Other regulatory laws followed, such as the *Pure Food and Drug Act* (1906) that provided for federal inspections of food products, meat in particular. At the time meat companies were selling adulterated meat, and drug companies poisonous patent medicines.[34]

Corporations and businesses have benefitted from the public largesse—from the Internet, cyber-security innovations, public roads and transportation systems, from a whole host of public investments—are now finding new ways to escape paying taxes to benefit the commonweal, or to find new ways to make money even if it threatens public safety.

The current American fiscal and economic debacle is symptomatic of a larger part of political dysfunction, the division of the country into federalists, who believe that government should work on behalf of the people, and anti-government advocates, or libertarians and their sympathizers, who oppose anything except war or military enhancement. It is a condition Francis Fukuyama describes as "America in Decay, The Sources of Political Dysfunction."[35]

Conservative political officials and commentators never fault corporations for malfeasance, but government itself as the instrument for economic crises. They focus on the threat to the dollar, on federal debt, on the number of people on federal subsidies, on foreign ownership of the dollar, and similar issues. The U.S. does have a debt imbalance but so does the rest of the developed world. The financial crash of 2008 affected all global economies, not just the U.S.

According to the General Accounting Office, the federal debt will reach its historic highs and unsustainable limits sometime around 2030, unless there are major policy or legislative changes.[36] Local and state debts show the same upward, and worrisome trajectories.

CONCLUSION

Today in contemporary America, the re-emergence of anti-government hysteria, born during the 300 years of American slavery and a subsequent century of southern segregation, misguided ideas about the complete freedom of market free enterprise, and reckless banking speculations, have castrated the American financial system and paralyzed government with venomous legislative dysfunction. Popular opinion is that the free disposal of property, and the unlimited exploitation of economic opportunities as rights are absolute and unconditioned by any restraint.

Capitalism as a functioning economic system fixes on wealth as the sole objective, on the exercise of the presumed right to pursue personal or corporate self-interest to the detriment of social goals benefitting all. Capitalism has become an end and not the means of achieving social progress. This does not imply that rigid socialism, such as that preached by Karl Marx, or that existed in the former USSR, when government owned all production, should be the model. Rather, an equitable partnership between government and business for the amelioration of society should be the norm.

Government equalizes business gains by taxing its profits to benefit the commonweal. Government relinquishes its social responsibility when it lowers taxes on wealth and thus exacerbates the inequality between wealth and the less wealthy who become increasingly impoverished.

Additionally, globalization, the loss of manufacturing jobs, the ever-present conditions of inadequate housing, dysfunctional schools, crime-filled neighborhoods, and a host of social disorders can also be cited as potential causes of American ennui.

Wage-earners and day laborers, middle management and small businesses, and householders and housewives, are all trapped in a stagnant economy from which recovery will be slow and uncomfortable. Questionable political and economic policies, and injudicious personal spending habits, have resulted in a depleted economy and declining tax funding for essential and long-term public service needs, including education.

What we do to ourselves is equally uncomfortable as what is done to us economically. We are in financial trouble, sometimes spending inordinately, and holding disproportionate amounts of personal debt. To future generations' obsession with money, let's hope that it may appear to be as pitiable as the 17th century's obsession with religious quarrels.[37]

As elevated primates, we have to rearrange our scale of individual and group values. Our national and personal economic interests must become a part of our lifestyle of interests, not the whole of its pursuits, our emphasis on finances subordinated to our social purposes. The discharge of dutiful obligations to society as a whole should never yield to the principle of acquiring wealth at any cost, or to governmental collusion to defer to corporate and business interests to the disservice and handicapping of a balanced social order. We are trustees for each other, and need to demonstrate that we really are improved beyond non-human primates.

Chapter 8

GOVERNMENT AS IMPERATIVE

I say to you that I have nothing but contempt for the kind of Governor who is afraid to follow the course that he knows is best for the State, and as for the man who sets private friendship above the public welfare—I have no use for him.

(Creon in *Antigone* by Sophocles 496-406 BCE)

Primates and other mammals collect around a leader, one who has proven worthiness in a battle or contest. Much of human history followed this primate pattern, to be governed by tribal chieftains, clan leaders, elderly wise men, kings and queens, and eventually presidents. Sovereignty of the people is relatively new in human society and distinguishes us from other primates.

In this chapter and the following I explore government in general and America's particular form of democracy, its functioning and its peculiar social and leadership legacy in particular. Government has become the bête noire—that which we most dislike but can't seem to live without, and yet extol as the best form of governance, even while we have little control over its actual laws, policies, or judicial decisions. In a democracy, we entrust decisions to representatives, the same as a troop or tribal leader in other primate societies.

The role and functioning of government often generates excessive passions, not just on behalf of a belief or ideology but at what government represents—its legislation, its debt, its over-reach, its foreign or domestic policies. The typical emotions are frustration and anger, the same emotions existing in other primate and higher order animals.

The American electorate knows and feels the dysfunction of government with, among other topics, the postponement of retirement and the social security safety net that supports it, with lower taxes that favor the super-rich and corporations, with a system rigged not to protect pensions, not to value equality for all, not to aid in the promotion of more jobs (and to devalue those at the lower end of the wage scale), and to slash needed benefits for education and the needy. This too encourages more anger and frustration, the rawest primate emotions.

The rule of the people, the government's protection of individual liberties and freedoms, is not an automatic exercise in much of the world. Militant and extremist Islamic groups believe that only God's law is permissible and that democracy is a sham. A kleptocracy like Russia, a highly centralized socialist government like China, and military ruled states, restrict personal freedoms to maintain power. Segments of the organic world, like the queen bee or the lion in charge of the pride, also have centralized control. So although we tend to believe that the rule of the majority is the best form of government, it is not the preferred governance system in either the primate realm or parts of the organic world that live in herds or common colonies.

Democratic government, under the rubric of political science in this inquiry, needs to protect the rights of the individual, but also to support the natural liberty of all citizens, as Adam Smith declares in this quote from *The Wealth of Nations* (1776).

> *To restrain private people . . . is a manifest violation of that natural liberty which it is the proper business of law not to infringe, but to support. Such regulations may, no doubt, be considered as in some respects a violation of natural liberty. But those exertions of the natural liberty of a few individuals, which might endanger the security of the whole society, are, and ought to be, restrained by the laws of all governments, of the most free as well as of the most despotical.*[1]

John Locke in *The Second Treatise on Civil Government* (1690), and Jean Jacques Rousseau in *The Social Contract* (1762), both advocated for a social contract among people for the formation and administration of a democratic government, while both were living under established monarchies. Both of these works served as the principal philosophical ideas for the classic American political documents.

All human primates crave liberty—within limits to do as they please—but also to organize themselves into orderly communities that serve to benefit all the people, and also to follow the appointed or elected leader. Not all humans have advanced beyond the Alpha male who serves as the chief, the tribal clan, or military leader, as is still the government example in Arab countries. Monarchies throughout the world are contemporary remnants of the power of hereditary kings and queens, and tribal chiefs and warrior clan leaders prior to nation states.

Which government is best for its citizens, and how should it function? You could spend a career investigating political theory through the writings of Plato, Aristotle, Cicero, Machiavelli, Montesquieu, Rousseau, Locke, Hume, Smith, and any number of less well-known philosophers and theorists of the state. The modern era has chosen the collective will of the people, or democracy (the rule of the people in Greek), as the principle from which a government should be formed.

If you believe the ideal state should provide sufficient goods to allow for the individual pursuit of happiness and virtue, then you agree with Aristotle. The ideal of political theorists, though, has rarely been realized in historical monarchies, tyrannies, oligarchies, aristocracies, or even some democracies, and certainly not in the semi-anarchist countries where tribal violence is the norm of governance.

In America, the tendency of a minority of individuals known as Libertarians is to have limited government, which is less a form of government than a belief in excessive intrusion, or Big Government, on individual liberties. Upholding a doctrine of free will, first conceived in the Enlightenment, promotes human liberty and has as little government interference in people's lives as possible.

Governments can reinvent themselves through revolutions, as America, France and Russia did, or through civil war, as America, England and Russia did, can depose monarchs, tyrants and dictators, and choose a new path for governance. On the other hand, violent revolutions spark such strong antagonisms when they overthrow governments that demagogues (or the military) rise to claim governmental control and authority, as did

Cromwell, Napoleon, Lenin, Mussolini and Hitler in Europe and military dictators in various parts of the world.

Those who grab power, according to Niccolo Machiavelli (1469-1527) in *The Prince* create enemies that have to be eliminated. To hold power they must create fear of retaliation to prevent revenge. The murder of dissidents thus occurs frequently to prevent rivals who challenge the power of the dictator.

A well-regulated government can certainly advance the happiness of its citizens by protecting it against invasions and external threats. But can people find happiness absent a functioning government? Monks, hermits and yoga masters can find happiness through contemplation, whatever the form of government where they live. But the majority of citizens cannot thrive successfully in political instability, anarchy or a slave economy. Governments like America's may not always be completely successful in attempts to bring justice and management to society, but neither should they be condemned to failure, as is often true in the cheap, cynical anti-government statements and sentiments of elected officials.

Government is frequently contrasted with the free market, or *laissez-faire* form of capitalism. However, a few institutions, regardless of whether well regulated or not, ought not to be able to endanger society. Bankers and financiers, car manufacturers, chemical companies, and petroleum empires quickly come to mind. Other corporations like pharmaceuticals, agriculture, and airlines control prices because they control products and fix prices for services or commodities. All this is fair enough in a global economy.

But if you believe corporations when they say that workers are their greatest resource—as they buy more robots to replace laborers, limit their wage earnings, and ship the manufacture of their product overseas—and that safety is their highest standard—when they can't recall enough defective ignition switches in their vehicles—then you should also question your uncritical judgment about deceptive advertising. Why should the business market be so free, as some claim, as to conduct business as they see fit without government interference, and not a democratic government that should also be free to impose regulations on deceptive practices that threaten the safety of citizens?

Most Americans want government to be efficient and reliable. They want a balanced budget because that creates a sense of stability. But they also want services—protection from terrorists or criminals, some compassion for the needy, the disabled, the elderly and children, a decent transportation system, reasonably good schools, and a healthy environment, among other desirables government offers.

Yet for all the controversy about government, its role and functions, its permanency and invasiveness, Americans are sadly ignorant of even the most basic of its guaranteed freedoms. According to a study by the Stats Group (and as reported by *USA Today 2015 State of the First Amendment* survey), Americans are not familiar with the First Amendment principles that define their freedoms. One-third cannot name a single freedom in the Constitution, and less than two-thirds were able to cite freedom of speech as one of the most significant of all individual protections. Less than 20% were able to cite the freedom of religion as one of the constitutional guarantees. Only 2% were able to identify the right to petition government for a redress of grievances.

What does this imply about American morale? For me it implies one of two things, but probably both: that we do not understand our democratic functioning very well, and that if we do its elected managers are incompetent in producing satisfactory results. The majority of the American public lacks basic civic understanding. According to the 2014 results of civic achievement by the National Association o Educational Progress (NAEP), only an average 23% of 12th grade students performed at or above a proficient level, and only 12% of Blacks. The distressed state of American educational achievement could not be more evident.

Identity and Tribalism

But beyond the traditional twin choices between government and business dominance, there is our national identity crisis. Our elevated attempt toward inclusiveness of all peoples runs headlong into our primitive fears of those dissimilar to us. As primates we are primed to fear those not typically a part of the community. Fear of others—of different languages, colors, identities—is a thoroughly primate response. Racism, discrimination alienation and similar responses are inbred in all primates as potential threats to the tight-knit social community. Heightened social consciousness that dispels these exclusionary attitudes is only a minority humanly acquired trait. It takes a sense of heightened social consciousness to rise above this primate fear to be accepting of those who are different in appearance, language or in a sub-culture.

Sometimes, a local government issue like the shooting of a African American teenager brings out underlying racial tensions. The political Left seeks color blind solutions to social problems, while the Right wants family and community values shared, forcing us to choose between the rights of the state in a policeman, over, for example, the value of a teenager's life. Neither idealistic vision holds. And that's because our shared history of race and its grievances is what ultimately defines us.

For example, a White police officer shoots a Black suspect in a crime or in an attempt to arrest him, even if the suspect is unarmed. Although not prompting as much media attention and large-scale demonstrations, the shooting in Phoenix of Rumain Brisbon, a Black father is representative of the conversation about police and the race of the victim.

On December 2, 2014, Phoenix police were asked to respond to a tip about a possible drug deal involving a Black SUV. Brisbon's SUV had stopped near an apartment complex as he was bringing home dinner from a fast food outlet. Police approached and he was asked to show his hands, a standard command when police approach suspects. Brisbon got out of the vehicle and continued to walk away, ignoring the command. Police then commanded him to get on the ground, but Brisbon again refused to comply, and yelled obscenities back at the police, according to witnesses. Police attempted to detain the suspect who resisted and a scuffle ensued inside the apartment. The officer saw Brisbon fumble for something in his pocket and, thinking it might be a gun, shot him twice in the torso, killing him. The pocketed item turned out to be a pill bottle. But a semiautomatic gun and a jar of marijuana were found in his SUV. Brisbon had previously pleaded guilty to burglary, had a DUI arrest, and had been arrested for possession of illegal drugs.

For Whites, such a shooting may appear to have been justified, especially if the suspect ignored police warnings, or if the policeman shot in what he considered to be self-defense. For many Blacks, such an incident is almost always about race, even if the victim is unarmed, and rarely about the victim's past criminal behavior. Whites generally want all the facts brought before a grand jury, a representative panel of the community just as a jury is in a trial, to determine whether or not a jury trial is necessary. Blacks always want the policeman indicted for an illegal shooting. But if a grand jury fails to indict, as they almost always do for lack of compelling evidence, protests occur, but just as often inexcusable rioting and looting results.

Police daily confront the schizophrenic, the pathological, the stoned on various hallucinogens, some whacked out on PCP, the drunk, the mentally challenged, and consummate liars about what they have done for police for stopping them. All detained are in many ways victims of regrettable circumstances, and all are usually the less fortunate in society, members of the lower socio-economic classes. Moreover, the police, also generally members of the lower middle class, are not themselves absolved of class and racial prejudice.

Police will detain individuals typically using handcuffs to control the situation. If a suspect is at a distance, police order the suspect to lie face-first on the ground, a form of humiliation, yes, but also a form of control so the suspect doesn't draw a weapon and can't flee. If just one policeman is available, the suspect is ordered to stay on the ground until backup arrives to place handcuffs on the suspect, again not necessarily under arrest but detained for questioning. Police too have been shot by suspects and are themselves fearful in situations where it isn't clear whether a suspect has a weapon or not, whether or not accomplices are nearby, or what his intentions are.

The social dilemma of a majority of police as White, the failure of the Black communities to become recruited into law enforcement proportional to their local populations, and the mix of social and racial classes is inherently inflammatory. Blacks sometimes view the dominance of White officials as a new form of superiority and dominance, even though Blacks are not always able to control young Black lawbreakers in their communities. White police traditionally view young Black males as suspects or potential lawbreakers, and themselves as possible shooting victims in retaliation for even detaining them.

The pairing of law enforcement personnel carrying lethal weapons, and the young Black males feeling threatened by police harassment has been symbolic of America's racial tensions for hundreds of years. This situation defines our history, highlights our social fears, and dictates our behaviors and attitudes. One way to understand these situations sociologically is that the sub-group, whether racial or occupational, protects its members and the inherent fear of those who are different rules primate instincts and emotions.

But there is more to our understanding of how government operates in our democratic society than how it confronts local racial incidents.

Election Choices: Choosing the Next Representative

Our guardians, setting aside every other business, are to dedicate themselves wholly to the maintenance of freedom in the State, making this their craft, and engaging in no other work that does not bear on this end.

(Plato, *Republic*, *3:395 b*)

The political spectrum in America divides fairly evenly between those who believe there should be government interference when issues of national import occur, such as legislative regulations on the banking industry, and those who believe in a *laissez-faire* form of democratic society in which government plays a minimal role in public life. These latter, mostly Libertarians and conservative Republicans, believe that private enterprise should be allowed to operate without, or with minimal, government restrictions.

The case for privatization of government is compelling. Public services offered for private operation might result in lower costs and greater efficiency. On the other hand, an inefficient government operation replaced by an even more inefficient private monopoly can be more expensive, wasteful and lacking in accountability or responsibility for serving the public good. Private contractors can be venal. The profit motive consistently trumps public interest. Shareholders and executives benefit at public expense, while public services can deteriorate.

The movement to privatize public services has gained momentum and the public is worse off because of it. Industry and government have to be partners in administering services, and businesses and corporations cannot be allowed to swallow the essential services only government should manage, such as schooling and prisons. Adam Smith in *Wealth of Nations* (1776) described these as: defense, post office, police, firefighting, public works, health, education, justice, transportation, banking, controlling monopolies, enforcing contracts and caring for the poor and infirm. There are multiple examples of how the public interest has become privatized and has influenced the public interest deleteriously. Here are a few examples.

Privatizing the mentally ill now means the public pays more in prison costs—often also privatized—and the majority of the mentally unstable are simply homeless.

American health services—except for the elderly under Medicare, the poor under Medicaid, and veterans—are dominated by the corporations of Big Pharma, Big Hospitals, and Big Health Insurance, not by Big Government. Health profits exceed health costs in the developed world, and insurance companies routinely deny coverage to those whom they deem unprofitable.

The Defense Department hires contractors to build military equipment that have perpetual cost-over-runs and little or no accountability. Half of all federal discretionary spending is allocated to defense.

In the meantime, essential public services like the police, firemen and teachers have had their salaries slashed as state budgets have run large deficits. The quality of civic life has declined and public servants have been roundly chastised for presumed lack of excellence, or as victims of state prolificacy spending or excessive tax cuts.

Privatization of public schools, a conservative movement to divorce schooling from public scrutiny and accountability with for-profit charter schools and voucher plans, has not shown consistent improvement in educational achievement.

Corporations, in collusion with legislators, have privatized public services so they won't be taxed for having them. The erosion of the tax base has occurred because of the reduction of taxes on the wealthy—starting in the Reagan Administration and again during the second Bush Administration—an erroneous policy statement that keeps getting repeated as if it had no effect on the economy, when in fact its unequal effects have been ruinous to revenues. The mantra that drastically reducing taxes on the wealthy will result in a stimulus to the economy has not produced the desired effects, and instead has further depleted a weakened economy.

About half of conservative American politicians believe that the role of government is to keep reducing itself, sometimes to the point of irrelevance, but is chiefly about cutting social service benefits and environmental regulations on behalf of public safety. This shibboleth is a concept misaligned with all other developed countries, and, as I will emphasize throughout this chapter, a cultural holdover from the days of southern intransigence about anything the federal government did to upset its peculiar way of life. Yet under the guise of freedom, religion and outright persecution—of gays particularly—these radical conservatives indulge in bigotry.

The other half, generally more liberal politicians, believe the nation under the umbrella and constitutional idea of promoting the general welfare, should extend government revenues and tax breaks to the less economically fortunate. Tax incentives like the extension of employer health benefits, mortgage interest, charitable deductions, earned-income tax credit for the working poor, deductions for higher education, and similar tax breaks, chiefly benefit higher income wage earners, and do not contribute to the reduction of income inequality. According to the Congressional Budget Office, the richest 20% of Americans receive more than 50% of the allowable tax breaks.

Actual social expenditures for social insurance are much less in the U.S. than in France, Spain, Germany, Italy, Denmark, Britain, Sweden, Portugal, Japan or Ireland, in that order. But income inequality, as is clear from the chapter on economics, is much higher in the U.S. than in any OECD country. Rather than just finding ways to tax the rich to equalize the palpable inequality, it might also be equally effective to minimize the tax breaks of the wealthy and thus allow more of the tax revenue to be distributed to those most in need.

All variables being equal—though they rarely are—in general there should be a healthy balance between allowing corporations and businesses to function more or less freely so that the economy grows and new workers can be hired, and the regulation of public health and safety standards from preventing businesses to place profits above the public good, and not to contaminate the soil, water or air to the detriment of society's safety.

Elected politicians are masters at winning elections, pandering mostly to local constituents whose biases they represent. Their speeches, whether on the campaign stump or state houses or in Congress, are paradigms of prosaic platitudes. We want them to speak in pithy, punch line slogans that will fit on a bumper sticker because we probably wouldn't understand the more profound explanations of how complicated government's domestic or foreign policy really is, and how it is managed.

Congressional representatives were once more effective at shutting down government than in managing it. The damage done so far, by failing to compromise on shutting down the government by not passing a budget and by indiscriminately slashing across the board budget reductions (known as the sequester) is a symptom of America's longstanding political dysfunction and its cultural and civilized erosion.

> *But when the guardians of the laws and of the government are only seeming and not real guardians, then see how they turn the State upside down . . .*
>
> (Plato, *Republic*, 4:421b)

Members of Congress are elected to serve the needs of the people not to hoist an ideological flag to their biases against the president or the government itself. But such are the acid sentiments of half the American population, they represent a flawed opinion about our democracy and the need to stifle more bellicose beliefs and serve all the people.

Although winning officeholders are quick to proclaim that, "the American people have spoken" when they announce their legislative intentions, the figures reveal that they are not as representative of America as they think. In the 2014 midterm elections, for example, according to *The New Yorker*, the House was composed of 89% of white Americans. The U.S. population was 51% women but only 20% of the U.S. Senate. White Americans were 63% of the U.S. population but comprised 90% of all federal, state and local officeholders.[2]

From the time Obama took office in 2008, no other American president has been so belittled and harassed, from the relentless, theatrical and pointless votes to overturn the *Affordable Care Act*, to the innumerable hearings on the Benghazi tragedy to embarrass the Administration, deliberately endangered the economy by refusing to raise the debt ceiling, to the prolonged and pitiful discussion about Obama's birthplace. (Oddly, when he was running for president no one ever questioned the birthplace of John McCain, who actually was born in Panama).

It was obvious to me and scores of Americans that these pathetic activities were meant to cast doubt on Obama's cautious leadership and his credibility when the rest of the world was marveling that America had elected a man of color, winner of the Nobel Peace Prize, and had seemingly moved beyond overt racism. Obviously, and to the chagrin of optimistic Americans, it had not. Radicalized Republicans and their rabid constituents let the government repeatedly grind to a halt because they are enslaved in anti-government ideologies, and find empathy for anyone not of their persuasion intolerable. This is an example of primate fear and ostracism at its worst.

The other strong primate instinct is religious identification, the exclusion of the group members not of the same religion, and who equate religious identity with nationalism.

Religiously fundamentalists believe the moral crises of the age are in gay marriages, the prevention of women to obtain legal abortions, and inserting religious liberty issues into the public domain. Their outrage is misplaced. The real moral crises is in insider stock trading, the obscenity of CEO pay, unwillingness to raise the minimum wages of working Americans to a reasonable level, the brazenness of banks to gamble investor funds in risky adventures, and the incestuous monetary relationship between legislators and corporations.

True, as former Secretary of Defense and CIA Director Leon Panetta suggests, Obama as president acted more like a law professor than an energetic leader on occasions.[3] Not every president can act decisively and recklessly by invading a country on false evidence as Bush did. But to his credit Obama addressed important issues with the economy, brought down interest rates, saw to it that the Congress passed a health care law, got the U.S. out of two wars, helped wage an effective campaign against terrorism, killed Osama bin Laden, and secured a nuclear arms agreement with Iran.

Confidence in all three branches of the U.S. government has plummeted, according to the Gallup Poll in July 2014, reaching a six-year low for the presidency (29%) and all-time lows for the Supreme Court (30%) and Congress (7%). Yet in the midterm election of 2014 only 33% of Americans voted, less than voted in a staged TV show. Americans have an opinion about the branches of government—if they understand the differences between them at all—but fail to exercise their democratic duty and vote their preferences.

Citizens choose not to vote (if they even recognize when it is time to vote), and vote a party ticket, and then often vote illogically when they are opposed to the very policies that their political party supports.

Granted, since people work in government, it too has worker errors, if not actual meltdowns in competence. The signature health care law literally melted down when its website repeatedly crashed after passage of the law. The IRS lost a bushel of Emails. A traitor of the NSA ran off with untold numbers of secret documents. The Department of Homeland

Security seemed unable to keep track of foreign students in the U.S. on visas. The Department of Veterans Affairs was unable to adequately take care of veterans. The Secret Service committed multiple blunders in protecting the White House from shootings at the building and invasion by an unstable intruder with a knife. All this was just in a two-year period in 2013 and 2014. Bureaucratic incompetence, not just in governance, roiled the concerned public.

THE DIVIDED HOUSE: CIVIL WAR, CIVIL DISCORD

> *Let us note among the guardians those who in their whole life show the greatest eagerness to do what is good for their country, and the greatest repugnance to do what is against her interests . . . And they will have to be watched at every age in order that we may see whether they preserve their resolution, and never, under the influence of either force or enchantment, forget or cast off their sense of duty to the State.*
>
> (Plato, *Republic*, 3: 412 e)

The Mississippi River divides the country geographically east and west. Slavery and its supporting states once divided the country north and south. We are not just a nation of two bickering, taunting political parties, only divided over domestic policymaking. We are two co-existing national mentalities, oppositional ideological differences defined by our turbulent history, and have been since the former colonies merged in 1776 under The *Articles of Confederation.*[4]

It was the *Declaration of Independence* in 1776 that renamed the "United Colonies" under British rule into "Free and Independent States." But it was only later that year under the rubric of *The Articles of Confederation* that the newly renamed United States acquired a more formal contractual authority under a confederacy. It took another decade before a truly national identity was constructed under the *U.S. Constitution.*

Constitutional framers in the summer of 1787 could not disturb slavery, as it was the equivalent of Banquo's ghost in *Macbeth* hovering over deliberations in the Philadelphia hall. The Constitution would never have been agreed upon by southern slave-holding states had slavery been amended or repudiated, so it was effectively postponed. A civil war did not erase the culture of White Supremacy or the Confederate flag that symbolized that southern culture.

The chief political division in the nation today is the rising tide of anti-government sentiment, a sectarianism that has carried over from the days of slavery and the century of segregation after the Civil War. According to former Governor of Massachusetts and presidential candidate Mitt Romney, the greatest threat to economic prosperity is not the budget deficit, the competition with China, the collapse of the housing market and the middle class, out-of-control entitlement programs, or the European debt crisis but "a government that is overbearing, intrusive, and demanding."

Anti-government advocates curiously never identify whether this applies to all three branches of government, or just one. The whole of democratic government is broad-brushed with a universal enmity that overlooks the constitutional fact that each branch has its own built-in checks and balances. Disgracefully, politicians pander to a public largely uniformed about the functions of its government, a democratic public that requires civic knowledge and involvement. Typically, politicians run a campaign against "Washington," as if the capitol city was the enemy.

Slavery and segregation hung like Spanish moss over the South trying to cover national shame but not resident pride. It's easy to understand an anti-government perspective in North Korea, Cuba, Burma, Libya, maybe even Saudi Arabia and Russia, but more difficult in America since so many immigrants come here because of American government freedoms and opportunities. The anti-government movement, of which the Tea Party and anti-tax groups form a part, had morphed from those who once saw the British government as a problem to now seeing constitutional government as an enemy, a dangerous transformation in a democracy.

The anti-government movement as a longstanding American historical phenomenon manifests itself in today's political debate as if coming from the political Left. In the past the anti-government movement came from the South. Today it is a southern mentality, now existing everywhere, that has become resurgent, polarizing the country. The same issues that divided us, absent slavery, in previous centuries are now palpably present in our politics. If we substitute today's political Left for yesterday's North, and the South for the far Right, then the contemporary terms fit a similar pattern.

I have concluded that the southern mentality has not been converted into American unity. The nation is knitted loosely together with thin strands holding the fabric tenuously and cosmetically. The South has never really broken free from its traditional anti-government attitude. United States history, and attendant presidential election campaigns, are a litany of compromises and indulgences to this inherent hostility. America cannot continue to exist with this permanent, internal discord, with the last vestiges of a long-ago civil conflict if it wants to achieve competitiveness in the global world. This is not about just a conservative counterbalance to a progressive, liberal agenda but a defined antagonism to a separatist way of life, notwithstanding the final removal in South Carolina of the Confederate battle flag in 2015.

I use ultra or radical Republicans to describe far right Republicans who promote this anti-government movement. They almost universally favor corporations and businesses, decry domestic programs for the less fortunate, block domestic government initiatives and tax increases even for natural disasters. From 2008 to 2012 to be a Republican politician meant you had to have a dogmatic catechism of denouncing Big Government, decry new taxes, be in favor of cutting federal programs (except defense, and without alienating those receiving Medicare or Social Security), and censuring President Obama for instituting a socialist federal health care program, even though it passed Congress. Hating government and being against it, and having no practical solutions for modern problems, are not attributes for actually legislating or administering government.

I am not contending that a strong conservative viewpoint is antithetical to American democracy, but that its more radical side is historically a part of a southern politics from an earlier era, and that its political habits have hardened. And although this is an indictment of sorts against the ultra-conservative bias of the Republican Party, it is not meant to be an implied endorsement of Democrats either. The ambition of both parties to gain power to control their political agendas is the overriding cause of all political activity.

That influence still resonates in our national discourse, political campaigns and legislative activities. The separatism that used to run only in the South has thoroughly permeated American society and fractured the political process so that little of electoral or legislative consequence gets accomplished. The principal political action of radical conservatives is to struggle against federal legislation, subvert democratic institutions like unions, belittle public servants whose job it is to enforce government regulations, and keep legislation from interfering with business and corporations or from raising taxes. Aggressive antagonism toward government is distorting politics with a resurgent and disquieting legacy born and bred in the South.

The contemporary oppositional and casual language we use to define ourselves—liberal v. conservative, progressivism v. capitalism, Labor v. Business, New Deal v. Free Markets—masks a long historical tradition of deeper social and political divisions. The North/South chasm has always been the nation's most pronounced distinction, and it contributes yet to our profound political disagreement. We are mired in a political dogmatism that rivals religious fundamentalism as a national characteristic.

The chief political division in the nation today has deeper roots than the fiscal debacle or the debt ceiling. Anti-government sentiment and White supremacy expose a divisive political sectarianism carried over from the days of slavery, a destructive and bloody civil war, and the century of segregation after the war. The political divide we experience today is a direct result of this historical legacy.

Since the ratification by the states of the *U.S. Constitution*, only people have rights, not states. Under the 10th Amendment, states have residual powers, not rights.[5] Today, the anti-government movement is ubiquitous, a southern mentality that is resurgent, polarizing the country again as it has from its earliest years. Eight of the first twelve presidents—Washington, Jefferson, Madison, Monroe, Jackson, Tyler, Polk, and Taylor—were slave owners. Two distinct and incompatible visions of America, indeed different civilizations, were allowed to live side by side, politically and economically, enduring multiple governmental compromises to perpetuate dual national existences.

The plantation economy was not a stand-alone economy, but in many ways had cooperative financial arrangements with the North, a form of collusion that today would violate federal laws. Southern plantation owners borrowed large amounts of money from northern bankers for new lands using slaves as collateral. Southern states backed the new bonds with the state's credit. When the southern economy collapsed, southern states defaulted on the debt. Many planters fled to Texas, an independent republic until 1845, and a few to Brazil where they began slave plantations anew. Clothing companies in New York and elsewhere made garments from southern cotton and resold the manufacturing to Southern buyers. Northern capitalists cannot be totally excused from collaborating with southern economic interests based on slave labor.

Remnants of racism and bigotry run deep in American culture and taint the politics American citizens mirror. The contemporary political spectrum has become a mixture of racist, anti-intellectual, and extremist views bubbling up from its dormant past.[6] We may speak about the closing of the American mind, as so many authors have, but in many ways it was never open to begin with.

Intolerance and prejudice is a corollary of this anti-intellectual narrow-mindedness and provincialism and includes creationists, certain evangelical Christians, White supremacists, doomsday cultists, and selected members of Congress who represent such groups. It exists partly because of a lack of informed education and partly because of an implacable bias, and has seriously divided American political, economic and social life for centuries. Its divisiveness and ideological partiality—and never admitting error when facts or evidence don't support misguided ideology—has penetrated the political system and stalled necessary national and state legislation.

A strong conservative viewpoint, as I've noted, is not antithetical to American democracy, but its more radical side is historically a part of a hardened southern outlook. The principal political action of radical conservatives, the historical descendants of southern anti-government sentiments to subvert federal legislation for minorities, undermine democratic institutions like unions, prevents government regulations on businesses and corporations, and prevents legislation that raises taxes going for a variety of federally subsidized social services.

The American South is still different because of its history of engrained bigotry. It is as different as Basque and Catalan are from Spain, Corsica from France, or American Indian culture from urban life. Because the South had been a separatist region evolving a distinct lifestyle prior to and after the Civil War, and through the disastrous period of Reconstruction and throughout the 20th century, it has been unable entirely to shake free from its conflicted past. The nation attempted unsuccessfully to make civil rights laws until 1965. Ultra-conservatives seeking a return to the purest form of the *U.S. Constitution* must accept that the original document condoned slavery as a property right.

Ultra-Republicans have been so keen to focus on simplistic and obstructionist ideology, such as opposition to Obamacare (*The Patient Protection and Affordable Care Act)* and any gun regulation, that they have blinded themselves to the complexities of a global economic life. Health care plans and banning assault weapons are about saving lives and preserving a healthy and safe society.

In American democracy there will always be a balance between the government's right to provide social services to the people—the constitutional phrase is "promote the general welfare"—and the people's right to choose life courses without governmental interference.

The reasonable move is to cut spending first to reduce inflation and reduce any deficit, then to cut taxes if necessary. There is no sensible purpose to cut taxes first if spending continues unabated because the loss of revenue increases the deficit and a possible increase in inflation. When Congress, or any legislative body, refuses to cut spending, the only recourse is to raise taxes, a condition Republicans loath but often find necessary to embrace, as even Presidents George H. W. Bush and Ronald Reagan did eventually.

On the other hand, there will always be tension from the political party out of power toward the party in power. But the recent trend is a revival of rancid hostility and not just opposition. We cannot divest ourselves of our divided past any more than we can throw off our genetic ancestry and cultural heritage. The pain and stigma of slavery won't go away, even after a war nearly decades ago that ended it, because the arrogance, language and culture of White supremacy survives intact in the population and surfaces periodically. As Alexis De Tocqueville noted presciently in his classic *Democracy in America*, "I perceive slavery receding; the prejudice to which it has given birth is unmoving."

It's no longer just a question of incivility and discourteous outbursts during President Obama's State of the Union address in 2008. It's about active attempts to dismember civil liberties, even voting privileges. It's no longer about several states hinting that secession, again, might be necessary, or about embarrassing politicians whose oratory and dialogue doesn't rise above schoolyard yelling arguments. It's about an alienated people and a history that cannot abide that they were militarily forced to unite with northern, liberal states, and cannot stomach a non-white president.

This is no longer a fight between the blue and grey and the secession of eleven southern states in 1861, but a lingering acerbic intermission from actual military conflict about whether or not radical Republicans—ideological descendants of the South and its sympathizers—are serious about participating in unity government at all, and can ever fully divest themselves of a racist, discriminatory and estranged past. The legislative strategy is not to compromise to aid the American public, but cynically to block and prevent a non-White president from gaining any political advantage.

Though slavery has been eradicated in the western world, the overt motives of those who oppose immigrant labor are just as vitriolic. Who should wear the badge of an American citizenship identity—even a president who must produce his papers—and who should be allowed to work in the U.S., are patently contentious issues in the anti-immigrant debate.

Contemporary events are a reflection of historical values perpetrated over more than two centuries of social and political division. America has never been thoroughly united, and its political compromises and temporizing over slavery for so many decades have inured generations to the mire of anti-government sentiment that contaminates today's national debates. Despite a ponderous debt and spending debate, we are a long way from overcoming our tragic social legacy.

Despite the current legislative impasse, the future will help defeat the reign of southern intransigence. The defeat will come from future demographics: an upcoming generation that will have broken from the southern mentality of White supremacy and anti-government sentiment, and from an array of Hispanic, African and Asian American population majority midway through the 21st century that together will overcome a plantation and segregationist mindset.

ANTI-GOVERNMENT STATE LEGISLATION

One example of pervasive anti-government sentiment comes from Arizona. The Arizona State Legislature in 2014 proposed an Amendment to the Arizona Constitution that restricted the state use of personnel and state resources to what it believed to be unconstitutional federal actions. Here is a section of Proposition 122, a referendum and ballot initiative sent to voters to approve.

> *Permits the State to exercise its sovereign authority by restricting state and local government personnel and financial resources to purposes that are consistent with the Constitution of the United States.*

A couple of items in this proposition need basic civic clarification. First, the *U.S. Constitution* is nothing more or less than the foundation of the U.S. national government. Among its powers are the absolute authority of, not just the Constitution, but also the "laws of the United States" and decisions by the U.S. Supreme Court. Article VI declares: "*This Constitution, and the laws of the United States which shall be made in pursuance thereof . . . shall be the* ***supreme law of the land****, and the Judges in every state shall be bound thereby, anything in the Constitution* ***or laws of any State*** *to the contrary notwithstanding.*"

It's clear from Article VI that not only is the Constitution supreme, but also the laws that Congress passes. Article VI vitiates any state laws that attempt to bypass or ignore this constitutional declaration. Therefore, there can be no legitimacy about the so-called "sovereignty" of states. The sovereignty of states did exist under the *Articles of Confederation* (1776-1787) prior to the adoption and ratification of the Constitution that supplanted it. There cannot be competing sovereign states under the Constitution, nor can states secede,[7] nor decide on their own whether or not they will abide by federal actions.

States cannot decide whether or not federal actions are constitutional or not, and the 10th Amendment does not give them that authority, though many uninformed conservatives claim it does. Sovereignty of state governance was a southern political strategy employed to permit southern states to keep the national government from transgressing against slavery, over which our tragic civil war was fought after eleven southern states seceded.

The Arizona State Legislature had proposed a *prime facie* unconstitutional referendum for the people to vote on. Here is the argument that a sitting congressman used to favor this proposition, as cited in the General Election brochure: "In recent years, federal overreach and judicial activism have often become the domestic enemies of the United States Constitution."[8]

Informed citizens need to know what "federal overreach" means. Is it laws passed by Congress, which, according to Article VI, are a part of the supreme national authority? Is "judicial activism" the decision of federal judges "that shall extend in all cases," as Article III of the Constitution states? How is it that "federal overreach,"—apparently laws passed by Congress—and "judicial activism," apparently decisions made by federal courts, have become "enemies"—of the Constitution is unclear, because congressional laws and federal court decisions are *de facto* constitutional.

In fact, "federal overreach" and "judicial activism" are conservative code words that mask anti-government feelings, and, when examined in the light of the Constitution itself, are found to be specious as well as misleading to the public.

Anti-Government Ideology: Segregating Ourselves

Alexis De Tocqueville's insight in 1838 is still pertinent today.

> *It is difficult to make the people participate in the government, but it is still more difficult to supply them with experience, and to inspire them with the feelings that they need in order to govern well. I grant that the wishes of the democracy are capricious, its instruments rude, its laws imperfect.* [9]

Though divided in political ideology, Americans are united on economic individualism, of competition in the marketplace, and of capitalism as one of the essential survival qualities of the human primate. We are in status, if not in name, capitalists who value the sanctity of property in all its myriad forms, including financial and industrial capital. The government through all its history, presidencies and congresses has always labored to sustain the industrial capital of the country. A vision of America is its currency, its chief religion.

Depending on one's political perspective, everyone can offer examples supporting one or another political view, and the following enlivened debate is perennial and often contentious. The maintenance of the equilibrium between what many view as government meddling, and what others see as a redress of corporate or business abuses, can actually be found within government itself, in its self-regulating constitutional division of powers.

The balance is in the electoral process when citizens elect representatives to offices, in the legislative process that formulates laws, and in the authority of the judiciary to rectify cases to be litigated. In other words, contained within the government itself are the seeds of the reform to correct governmental or institutional imbalances to society, by means of responsible, newly elected leaders, new laws, or through the judiciary.

Historically, however, Americans have been split even further than the reasonableness about the role of government might appear. A legacy of our totally diverse economies, into north and south more or less geographically, but really into a manufacturing economy and a plantation economy with slave labor, has been our more persistent distinction. It is not just the economies themselves but perspectives on how Americans viewed individuals within the country, especially how southerners perceived the slaves and their descendants. Virulent anti-government sentiments have their origin in this geographic and cultural divide, our troubled history, and it exists still in our state and national politics. According to John Quincy Adams:

> *Slavery is, in all probability, the wedge that will ultimately split up this Union. It is the source of all the disaffection to it in both parts of the country.*[10]

Before the American Civil War nearly every issue impinged on, and revolved around, slavery. American Indian policy, the attempt by southern states to nullify congressional legislation, the buildup by southern states on the issue of states rights, taxes on manufactured goods, the role of the national government in public improvements, whether the national government should have a bank, the delivery of the mail over abolitionist literature, the imbroglio over Texas, the transformation of territories into states, even gag orders for speeches about slavery in the House of Representatives (the House passed a gag order denying members the right of free speech about slavery on January 28, 1840 by a vote of 114-108)—all these issues were related in one way or another with slavery. Southern interests in the mid-19th century controlled Congress. At the time, slavery triumphed over human rights, and its advocates overrode constitutional rights to defend it.

Outright racism, concealed bigotry, positive sentiments for segregation, and the election of America's first African American president, combined to form a toxic national cauldron. Such a mixture contributed to congressional dysfunction, as it did in 1840, and to the rise of radical militant groups during the Obama Administration from 2008-2016.

There is a residue of intolerant temper in the American psyche that is a part of the national mentality, exhibited not just in legislative dysfunction but in the society and culture itself. It is a symptom of a past that extols a customary but anachronistic rule of life that still contains elements of bigotry and fanaticism. George Packer writes about this in the "Unwinding" of America in his 2013 book with that title.[11]

But the country was never really knit up in a manner that was meant to sustain its unity, not with the explicit permission of slavery in the *U.S. Constitution*, and the unbalanced electoral advantage of southerners with the three-fifths voting franchise. Prior to 1860 southern slave states had an unfavorable election advantage. Those running for elected federal offices were able to count three-fifths of their slaves as voters, when slaves could not vote. Hence, all elected officers for U.S. president or members of the U.S. Senate or House of Representatives from slave states, because of this unfair constitutional anomaly, did not fully represent all male persons and were thus unrepresentative of the union. Of course at the time women could not vote either. Northern non-slave states elected officials from all male voters.

This odious compromise embedded was totally undemocratic. Slaves and women were counted as voting representatives but could not vote. Southern representatives were disproportionately representative even of the less populated South.[12]

Poverty and income stagnation for the middle class are also partially the result of political decisions in the U.S. Too many Americans, partly because of ignorance of civics and constitutional principles poorly taught in schools, only honor antagonism and harsh ideologies that define the U.S. heritage. The political parties today are poisonous and have gravitated towards respective conservative and liberal power bases instead of the common good. Campaign finances, the flood of corporate money to members of Congress, the so-called think tanks belching ideological attacks on the other party's ideas, and the trivialization and sensationalism of the electronic media have all contributed to a degradation of public discourse and the failure of a national commitment, leading to the solution of common problems.

The bankruptcy of Detroit and a few smaller cities should be a symbol and prediction of the future of urban America. The polarization of American society and the inability of its poor and eroding middle class to generate sufficient personal savings or taxes essential for public necessities foreshadow what urban America, once the great economic hope of civilization in the 20th century, might become in the 21st century.

The shutdown of Toledo's drinking water in the summer of 2014 was the direct result of an increase in potassium bled into rivers and into Lake Erie from widespread pesticide use. Actual pollution is occurring, not just the political pollution coming from the far political right.

The explosion in February 2015 at ExxonMobil's Torrance, California processing facility was a result of deliberate violations of the safety codes, according to the Occupational Safety and Health Administration. The petroleum company had been cited for 18 serious violations that endangered public safety and failed to correct the dangers and was fined $560,000. Not just accidents, but the failure of corporations to obey the laws and to persistently violate known safety procedures is reason enough to have tough regulations enforced vigorously.

Special interest groups lobby Washington for favors by generously giving money to re-election campaigns. When I worked in Washington in 1971 there were 171 registered lobbyists. By 2009, there were 13,700. Collectively, they spend about $3.5 billion annually advocating for their special treatment in legislation, or in stopping legislation they believe might damage their industry. The general public has little knowledge of these dealings, and thus is unrepresentative of the electorate's known wishes.[13]

By 2015, the court had ruled in favor of the *Affordable Care Act* and gay marriage, and support for the *Fair Housing Act*. However, prior to these positive judicial rulings it had limited women's right to control their bodies, and laborer rights to challenge wage discrimination and totally eviscerated campaign finance laws and the nation's *Voting Rights Act* that Congress had repeatedly renewed.

As a result of money considered legally as free speech—a rational stretch by itself—direct spending on state judicial elections had more than doubled, and special interest and partisan group spending on state judgeships has accelerated—a seven-fold increase since 2000. Judges have now had to become professional fundraisers if they want to remain in office. Can Americans be absolutely convinced that judicial decisions are fair and impartial if their decisions bear the marks of significant campaign contributors from special interest groups? Money pouring in the judiciary poses a significant threat to judicial independence.[14]

The radical wing of the Republican Party has not shown to be responsible advocates for reasonable economic practices, using only crude slogans about human liberty and antipathy toward government actions to thwart economic progress for the benefit and the "general welfare" of all citizens. Contrary to incessant radical Republican claims that the deficit needs to be reduced, it declined by half from 9% of Gross Domestic Product (GDP) in 2008 to just 4% by 2013. Also in late 2013, overall government spending fell by nearly 5%. Federal government spending dropped by 12.6%, partially caused by the government shutdown in 2013.

Contrarian politics, severe political dysfunction, considerable ignorance, and civil and cultural polarization, from which there is no easy recovery, have contributed as much to slow economic growth as have the lackluster and highly extolled free enterprise private sector.

Let me close this section with a note of current reality about who a person is, an example that comes from the U.S. Supreme Court. The normal understanding of a person is that of a human primate, flesh and bones and all that goes with that. One does not often think of a person as an economic entity, a corporation. Yet in the twisted legal logic of the majority of conservative members of the U.S. Supreme Court, that is exactly who a person is, with all the attendant privileges.

The U.S. Supreme Court ruled in January 2010 in the *Citizens United* case noted above that corporations are persons, and therefore have free speech rights and can give as much as they like to political campaigns. One wag observed that he would believe corporations are persons when Texas executes one.

Let's use a bit of whimsical logic. Question: were corporations spontaneously generated, the same as germs which were once thought to be created? Or were they sexual reproduced from humans? Inquiring minds need to know because it appears on the face of it that no biology was involved. Making a contractual entity a biological entity is a serious disservice to biology, nature and science, though apparently not the law, and certainly not in the minds of a majority of the court of highest law.

And then if, as claimed by the court, corporations are persons, what is their gender? Persons have to be biologically male or female. Without this legal equivocation decided, they cannot simply be neuter people and cannot duplicate themselves.

To maintain its twisted logic, the U.S. Supreme Court, having ruled that corporations are persons, held on June 30, 2014 that a business, Hobby Lobby, a retailer of arts and crafts, can exercise its religious freedom by withholding health benefits from its employees because of the religious beliefs of its owners. I guess if the business owner were a Muslim, it would be legal, exercising his religious right, to declare jihad against his workers for not being productive enough.

Undecided is whether or not this applies to the business itself or just the owners. For example, if Hobby Lobby is sold, will be new owners be under the same restriction for the business as a person, or can they decide to impose no religious persuasion on its workers and in effect reverse this decision? In the court's strange logic, the rights of workers are subservient to the religion of the owners. Marx would definitely not approve that businesses can be people, but neither would most thinking citizens.

For two and a half millennia the American South enslaved African Americans, bought and sold them as property, and thus denied them any rights. For those who claim that a fetus is a human with rights at conception, here is a counter-argument: fetuses rely exclusively on their lives to the women where they temporarily reside. They are a woman's property, and therefore have no rights until she releases them from a natural bondage. The fetus is a slave to the mother's body. Those who grant a fetus individual property rights before they are technically and legally persons are basing their argument almost always on religion, and accord no legitimacy to science, logic, law or the history of entitlement. Such is the triumph of evangelical religion, and its political adherents, over the advances to civilization over the past three hundred years.

Struggles with the role and function of government are a part of democratic discussions. But the turbulent history of our peculiar form of partisan government and dual economies over nearly three centuries have burdened us both with a wounded society and a flawed Congress, sadly maladjusted to favor an antiquarian ideology against the poorer classes, who happen to be mostly ethnically different. Society's imperfections are reflected in legislative and judicial defects. We cannot find a ready cure to our governmental weaknesses until we remedy our own social flaws.

Cannons at Shiloh: The Civil War Still Resonates.

Chapter 9

PRESIDENTS AS ALPHA MALES

In summing up this abbreviated articulation of who we are as both participants and custodians of democracy, let me highlight two presidents and their policy initiatives. We have a unique opportunity to examine our elected leaders, using my theme of primate terminology and biological origins, as they act like tribal chiefs and national Alpha group leaders. The analogy is not just a bit of whimsy but a call of honor to our primate ancestral heritage, a sort of badge of respect to the protection of the species and not a dishonor or disservice to presidential abilities, or to non-primates.

As the Alpha male, whether winner of the contest against other males, or winner by the electorate who believes you are the best of the candidates to lead the country, a president has the responsibility of safeguarding the citizens, a primary constitutional obligation. This requires a reasonably strong aggressive tendency on the part of the Alpha male, exhibited to both citizens and potential challengers or contenders.

War is the extended national version of a cockfight, an elk or ram battle with horns, or a duel among rivals. War is the primate aggression among humans to protect and defend members of the same nation, or to demonstrate the aggression of a president and his council under the pretense of defensive purposes. I don't want to describe this condition too metaphorically or thematically in the presentation of the following presidential prerogatives except to note that the national tendencies of aggression, including presidents as Alpha male protectors, do apply to foreign policy initiatives.

World War II was America's response to the protection of Europeans and to the Japanese attack at Pearl Harbor in Hawaii. The Korean War, the Cold War and Vietnam were responses to the perceived and aggressive enemy of global communism. The first Iraq war in 1991 was a reaction to the loss of needed oil from Kuwait. Afghanistan was a war in reaction to the attacks in 2011 by an Islamic militant group, al-Qaeda and not a specific country.

However, Afghanistan soon became the wrong country to fight the attackers of 9/11 in, as the journalist Carlotta Gall makes clear in *The Wrong Enemy, America in Afghanistan 2001-2014*. When Al-Qaeda moved to Pakistan, we should have moved there to eliminate them, as the Taliban in Afghanistan were not our original enemies. They were brutal Islamic militants but not the enemy that attacked us. Al-Qaeda was the enemy, and when the U.S. attacked Iraq, the Al-Qaeda leadership moved into Pakistan and re-grouped with Taliban fighters under the auspices and protection of the Pakistan military. How crazy and delusional is it to stay for fourteen years in the country that your enemy has fled, and then to fight a group there not your original enemy?

The American propensity to go to war is an illustration of America's view of itself as the Alpha country. A streak of militancy in a large part of the population feeds into that sense of superiority. Politicians tap into this national hubris, and the results usually cost the taxpayers heavily in funds and the loss of military youth.

American foreign policy is frequently an aggression to actual or perceived threats, often based on misinformation, lack of knowledge of the presumed adversary or country, and a reaction that is sometimes out of proportion to the threat—almost exactly the way it is in the non-primate populations whose Alpha males use their power to intimidate potential rivals and threats. Reagan's invasion of the tiny Caribbean island of Grenada in 1983 was an example of a power play to demonstrate his resolve against the influence of Cuban communism in the southern hemisphere. George Bush's invasion of Iraq in 2003 was similar, and likely waged to satisfy his vengeance for Saddam Hussein's attempt to assassinate his father in Kuwait. Neither of these wars directly threatened America or its citizens.

I'm choosing Franklin D. Roosevelt (FDR) the Democrat, and Ronald Reagan the Republican, as Alpha male presidential leaders. Each was transformative in political national policy. Both FDR and Reagan gave impetus to social changes and political policy that remain with us in domestic initiatives. In keeping with the theme of this book, I see them in the same way any primate or higher organism views its leader: as representative of the species and the protector of the survival interests of the group, even though each came from distinct political parties and had opposing ideological views.

My twin choices are of course not meant to disparage or lessen the importance of other American presidents as leaders. I admired Harry Truman as a Midwestern exemplar of American courage and steadfastness. Dwight Eisenhower was born in 1890, the same year as my grandmother. I watched his presidential candidacy on the first TV the family owned in 1952, cast my first ever vote in a presidential election for him. In 1969 I watched on Pennsylvania Avenue as his funeral cortege passed by with his coffin followed by the riderless horse of a fallen military hero.

I was delighted with the short and regal presidency of John F. Kennedy, the aversion of the Cuban missile crisis, and sad national aftermath of his assassination. I came to work in Washington DC during the Johnson Administration and I lauded his determination and tenacity in facing down southern sympathizers, and in persuading Congress to pass civil and voting rights legislation and wide-ranging education laws. I worked for the Nixon Administration, and though I distrusted him from the beginning—a sentiment shared by the public as the Watergate scandal unraveled his Administration and his deceits, and he would have been impeached had he not resigned—he turned out to be a moderate in domestic policy and a genuine leader in foreign affairs.

So presidents come and go, but for a very few the legacy they leave behind remains and re-fashions the contemporary landscape. That's why I think Roosevelt and Reagan are a bit more relevant than others.

THE FRANKLIN ROOSEVELT LEGACY (1882-1945)

Franklin Roosevelt, born of wealth and privilege, was a political pragmatist, unafraid to change his mind. In 1922 he spoke out against government regulation of business. In 1929 speech when he was Governor of New York he decried the dangers of a government relationship with industry. In 1933 he did not favor the federal insurance of bank deposits. Congress created the *Federal Deposit Insurance Commission* (FDIC) anyway that guaranteed peoples' saving accounts. Each of these political stances would today mark him as a conscientious Republican. Yet for all his charm and beguiling personality, he was a practical man willing to experiment with policies, not an ideologue. And he did change his mind as the times warranted. Despite his Republican-leaning ideas, as president he promoted some of the most far-reaching domestic programs in the history of America. The history of his legacy is worth revisiting.

By 1936 the most popular song, that I learned to sing as a child, was *Pennies from Heaven*, an upbeat song and lyrics about clouds raining pennies "for you and me." It gave encouragement to everyone that the coin worth the least would start a financial recovery. I recall listening to his fireside chats in the warming comfort of my grandparents' kitchen. In his avuncular voice he explained in his first message of 14 1/2 minutes he spoke about how banks worked and how they failed. He said that: "hoarding has become an unfashionable pastime," and encouraged people to put their money back into banks. And they did, re-depositing funds they had withdrawn in panic.

But the real bounce back from a Great Depression was not going to be limpid optimism but the role of the national government, which prior to Franklin Roosevelt had been passive and non-committal about peoples' problems. Suddenly, in Roosevelt's first hundred days in 1933, a flurry of bills were passed that demonstrated that help was on the way, and in a big way. Roosevelt would show that government would not be indifferent to the plight of the people. He would show that swift and strong action would help speed national recovery and secure the wellbeing of all citizens. And people believed him.

He created farm subsidies, bank regulations, relief programs and social security. One of the unemployed men who found work in the government's relief program known as the *Works Progress Administration* (WPA) in Illinois, acquired because he had campaigned for Roosevelt, was John Edward Reagan, or Jack, father of Ronald Reagan.[1]

After 1929 over 5,000 banks failed and 60 million lost their savings. Shortly after Roosevelt's inauguration in 1933, he began what came to be known as fireside chats.

The National Recovery Administration (NRA) passed by Congress was intended to set fair standards between business and labor by establishing codes for prices, production levels, working hours for laborers, and minimum wages. The NRA codes were found by the U.S. Supreme Court to be unconstitutional because they granted to the executive branch the power to legislate.

Almost immediately after his inauguration in 1933 he proposed the *Emergency Banking Act*, the *Civilian Conservation Corps* (CCC), the *Agricultural Adjustment Act* (AAA), the *Tennessee Valley Authority* (TVA) that served seven states with rural electrification. All of this legislation in 1933-35 put people to work and, as with the CCC, and gave them schooling as a part of their work assignments. Beginning in 1933 the CCC provided jobs for three million unemployed youth, and aided park and forest development, disaster relief efforts, control of soil erosion, wildlife protection and historical preservation—all programs the country needed to protect and preserve the national heritage. Its results are still visible today in our national parks and, for example, in cabins built by CCC employees along the Appalachian Trail.

Practical economic programs dictated Roosevelt's executive actions based on a re-evaluation of his own political values and his keen political acumen. Hence the New Deal—his umbrella term for all his programs—was not a coordinated set of strategic policies but a set of improvisations not anchored in any grand economic theory.[2]

By the time of undeclared naval war in 1939 against Germany, and the declared war in December 1941 with Japan and Germany, the New Deal had exhausted its influence and full-throttled manufacturing production revived the economy and put people to work. We should not automatically conclude from this that war is good for a flagging economy. What we can conclude is that a determined Roosevelt and a revived American economy and manufacturing base helped defeat the forces of tyranny and aggression from Germany and Japan from 1941-45.

If you have flown into or out of LaGuardia airport, driven through the Holland Tunnel, seen Bonneville or Grand Coolee dams, you have experienced the engineering results of the Roosevelt legacy. But one of the most lasting legacies did not just help citizens out of the poverty of the Depression. It remains as a cornerstone of social relief to millions—social security, intended as a supplement to retirement savings. Over the years since its adoption nearly every president has had an attempt at reforming it, especially in the years when outputs for recipients exceeded incoming revenue from workers.

Franklin Roosevelt's New Deal was an enhanced version of cousin Teddy Roosevelt's Square Deal. It was a series of government programs meant to lift desperate people from the thrall of deteriorating economic conditions. It brought back government regulation of the financial sector, and arrested speculation rage in risky stock equities that resulted in the Great Depression of 1929. By the time Franklin Roosevelt began his Administration in 1932, social security was just one part of a series of programs aimed at restoring the economic security of Americans, including emergency relief, housing and employment insurance.[3]

The earned income credits the retired receive known as Social Security, or the *Federal Insurance Contributions Act* (FICA, but originally known as the *Social Security Act of 1935*), is a regressive payroll tax imposed on all employees and employers that automatically becomes available at retirement for those who sign up. A beneficiary has to have worked for 40 quarters or ten years to receive benefits. It is not an entitlement, as so many claim, nor a personal savings tax. It is a hedge against those who don't save enough, or any, funds for retirement.

It was when enacted, and still is today, one of the boldest, most comprehensive pieces of social legislation in American history. Charged with the responsibility of protecting all Americans from financial calamity as they approached retirement, it has succeeded in enlarging the role of the national government by fulfilling the constitutional ambition of providing for the general welfare. Today the elderly retired are able to eke out a precarious living, courtesy of government intervention, that would have been impossible a hundred years ago, and kept them from being doomed to an impecunious existence.

In his State of the Union address on January 6, 1941, when the U.S. faced the dangers of Hitler's Third Reich and Japan's military expansion, Roosevelt noted how it was necessary for the Congress to help Americans build their domestic needs too.

> *We should bring more citizens under the coverage of old-age pensions and unemployment insurance. We should widen the opportunities for adequate medical care. We should plan a better system by which persons deserving or needing gainful employment may obtain it. I have called for personal sacrifice. I am assured of the willingness of almost all Americans to respond to that call. A part of the sacrifice means the payment of more money in taxes.*[4]

It is a message that appears to serve our present day circumstances as well too, though a growing number of Americans and their representatives cling to the belief that their sense of democratic liberty is opposed to such government action.

Roosevelt may have recognized the discrimination in the South, but the national economic depression and the war prevented him from initiating any sustained programs there. But he did recognize southern poverty. According to the presidential report, *The Economic Conditions of the South*, when Roosevelt was president during the Great Depression, the "South presents right now the Nation's No. 1 economic problem."[5] Rather than address civil rights directly, Roosevelt focused on the economy in a report that detailed how the South was economically disadvantaged. FDR did not believe that businesses should make all the essential decisions on society, or that government should pursue a policy of disinterest in the economy, or that popular government was dangerous—all policies advocated by the Tea Party movement. Roosevelt thought of solving problems, not about upholding political dogmas.

In 1938 there was a disproportionate number of adult workers to dependents in the South, as many workers were displaced by machines, notably tractors, and White workers were taking over jobs once reserved for Blacks, while the most skilled workers migrated to the North. The level of poverty was great throughout the nation but acute in the South. This could only deepen the ethnic divide through wage manipulation and taxes to support schools. The South was educating one-third of the nation's children with one-sixth of the school revenues. Teacher salaries were four times less than in the North. As a result, there was nearly four times the illiteracy in the South as elsewhere. Economic setbacks in jobs, schooling, health and housing are still the context for the South's coming to grips with its civil rights past.[6]

Roosevelt's opponents on the right were typically conservative in response to the New Deal, that flurry of legislation in 1933 that plunged the government headlong into the nation's economic recovery. Contemporary opponents of government intervention react just as strongly today to government's provisions in health care, or any government social programs.

The Swedish sociologist Gunnar Myrdal's classic study of southern race relations in *An American Dilemma* in 1944, and Stetson Kennedy's 1946 book *Southern Exposure*, a realistic critique of the undemocratic South, both laid bare the racist and economically underdeveloped underbelly of the nation.[7] After World War II the South still held a firm, White grip on politics and economics. States routinely disenfranchised Blacks and circumvented the 15th Amendment ratified in 1870 by requiring a poll tax or literacy test for Blacks to register to vote. Today, several Republican-controlled states have passed laws reducing early voting and legislated other voting restrictions, like requiring photo ID under the ruse of voter fraud, that tend to disenfranchise minority voters.

Roosevelt's leadership and legacy was not just in his persuasion of members of Congress to pass his legislative requests, or his administration of a vast complex of new domestic social institutions, but his inspiration to the citizenry to believe in themselves. In 1936 he won 68%

of the popular vote and 46 of the 48 states. (Alaska and Hawaii were admitted later as states). People did believe they had come out of a financial dilemma because of his initiatives and his calming manner and voice, and they did believe in him.[8]

The leadership of America during World War II cemented that relationship. That is why on the late morning of April 12, 1945, when the nun who was my elementary school principal came into my 5th grade classroom and announced that Roosevelt had died after 12 years in office, then abruptly left. He was the only president I had ever known. The whole class abandoned their lesson that day and had an audible collective cry, as did the rest of the country.

THE RONALD REAGAN LEGACY (1911-2004)

> *Dad was also a deeply, unabashedly religious man. But he never made the fatal mistake of so many politicians wearing his faith on his sleeve to gain political advantage. True, after he was shot and nearly killed early in his presidency, he came to believe that God had spared him in order that he might do good.*
>
> (Ron Reagan Jr., June 11, 2004)

I was living in Japan when Reagan gave his memorable speech nominating Barry Goldwater at the Cow Palace in Daly City, California in October 1964. It was so electric to conservatives that they immediately began a campaign for his presidential bid in 1968 even though he had never run for any elected office. Reagan had been a B grade actor in forgettable movies and a pitchman for a TV program sponsored by General Electric. But Republicans were so desperate for anyone that held a similar anti-government ideology and a dislike for taxes and deficits, they were eager to crown Reagan. Instead they nominated Richard Nixon who was the only president forced to resign in 1973 in disgrace because he would have been impeached had he remained in office.

Running against "Washington" and the over-reach of government was standard campaign fodder from both parties, and the slogan always received applause even though no individual, like the president, nor branch of government was mentioned as the cause of the disdain. And no one inquired why, if such contempt of national authority was evident, why anyone would seek to govern it.

Reagan was elected Governor of California in 1966 and re-elected in 1970. He was elected president in 1981 and re-elected in 1985. The influence of his policies and persona were profound.[9] During his 16 years in elected office he held to his dual strong views—anti-communism, and limited government. But he was also a practical politician and did raise taxes, eventually, and he compromised with Congress on deficit reduction.

The mid-1960s were tumultuous in American society as Reagan came into prominence as the twin movements of malaise over the war in Vietnam converged with the long overdue rectification of federal civil rights for African Americans. For over a hundred years, from roughly 1866 to 1970, southern states had disenfranchised Blacks and denied them equal rights as citizens, segregating them in schools, businesses and public facilities. The passage of the U.S. Supreme Court decision *Brown v Board of Education* in 1954 judiciously desegregated public schools, but not without willful and sustained resistance throughout the South. President Lyndon Johnson's legacy of persuading Congress to pass the *Civil Rights*

Act of 1964 and *The Voting Rights Act* of 1965 solidified in legislation what the South could not legislate on its own because of the entrenchment of White Supremacy as the ugly legacy of slavery.

The violence that erupted when Martin Luther King was assassinated in April and Robert Kennedy in June 1968 was a watershed moment of the built-up tension in the Black communities over the lack of civil rights and the fury at those who assassinated their heroes. The national schizophrenia, basically Black and White, we have towards each other burst into full-blown violence in the late 1960s. Non-human primates exact vengeance on members of the same species when threatened. Human primates are no different.

I was in Washington DC in July 1968 on my first government consultancy assignment when I witnessed jeeps with marines manning mounted machine guns patrolling the streets. I was trying to reconcile my enthusiastic experience in our nation's historic capitol with the level of violence and hatred exhibited by the rioting Black community. I realized then that social revolutions, like the American Revolution, sometimes do have to occur through violence. Detroit, among other cities, was burning and the city has never fully recovered. The hatred shown to President Barak Obama after he was elected is a taste of the residue of bigotry and racism still stirring in much of the public.

Reagan was born in Illinois and his father Jack, as I noted above, worked for one of FDR's democratic work projects. But son Ronald was a solid conservative at heart and he sided with southern sympathizers. But no national conservative leader was going to be elected in the 1960s because the public sought leadership from Johnson to end the war and resolve the nation's long history of southern intransigence to recognize Blacks as equal citizens.

Reagan didn't like the *Civil Rights Act* and he didn't like Medicare. He didn't like much of anything the federal government did except provide for the common defense. He gave many speeches decrying government enlargement but without explaining that the so-called encroachment of government was a democratic prerogative of elected representatives of the people.

Although he had sympathy for the less fortunate, he did not want government sponsoring programs that simply handed out money to them. Thus, he was in favor of promoting the general welfare, the other clause with defense in the Preamble of the Constitution, but not in favor of paying for it. He hated communism and raising taxes, his twin but impractical operating principles, although he did raise taxes twice, both while as governor and as president. He rarely administered or managed any governmental activities, leaving the details to others.

Reagan was the image of conservative America, naive about flexibility in government, and the enlargement of multiethnic social life in America, and the guiding principles of the fragile economy. Like most conservatives he believed in supply-side economics—cutting taxes and deficits—because it suited his ideology about wasteful government policies spent on America's poor and unfortunate. Hence, he rejected the Keynesian policies that favored demand-side economics that preferred government action and in the economy that stimulated growth during times of recession. Supply-side policies prevailed in the conservative Congress that dominated during the Obama Administration from 2008 to 2016.

Reagan's antipathy to government programs extended to all federal welfare programs but specifically to social security, a payroll tax imposed on employers and employees to fund future retirement funds. Shortly after the assassination attempt in 1981, he and his team attempted to cut the social security payroll tax. When the bill arrived in Congress it was immediately assailed and deleted. By 1983 after a bipartisan commission he established recommendations that included making previously exempt federal employees a part of the system, and incrementally increasing the retirement age. Congress approved the reform package.

No official thereafter dared to touch social security until George W. Bush as president again proposed privatizing it by placing the funds in individual equity accounts just before the great economic collapse in 2008. Like Reagan's first attempt at redefining social security, Bush's privatization scheme died stillborn.

Reagan aligned himself with antediluvian evangelicals—who offered only negative policy initiatives, like anti-abortion legislation and creationist ideology in schools—and with gun nuts, and anti-communist radicals, and anti-government social policies that used tax money to support the so-called welfare state. He followed traditional conservative thinking about the growth of government that he never applied to defense spending. That game began with his Secretary of Defense Caspar Weinberger who inflated the cost of military projects, and then agreed to cut them so it would appear that the defense budget was saving money. Reagan agreed with this charade and the public was never the wiser.

The founding fathers, Madison and Jefferson predominantly, did not favor a standing army and detested war. Reagan, ignoring or not knowing much history or anthropology about real or presumed enemies, enlarged the military so much that it took more than two decades for the nation to unburden itself of the fiscal damage done by bloated defense budgets.

Reagan had no ideological nuance in him, and had only a fervent hatred of communism. Russia (and Nicaragua, Cuba and Iran) was evil, Russia had designs on America, and the country must spend all its resources on countering its perverse ideology. He used scare tactics and deceit in speeches to delude the public into thinking America was all but under attack.[10] His Iran-Contra dealings were an impeachable offense. He never sought congressional debate or permission to go to war on Granada. He badly managed the marine bombing in Beirut in 1983 and its aftermath.

He had the CIA mine the harbors of Managua and never told Congress about it. Even Barry Goldwater, Chair of the Senate Intelligence Committee at the time, was furious. The Boland Amendment made it clear that Congress was in charge of declaring or making war, not the president. Miffed at this congressional rebuke—in actuality a constitutional prerogative—Reagan then enlisted private contractors to fund his secret war anyway, a program managed by Oliver North in the White House with Reagan's approval.

Then Reagan had Carter's hostage problem. Hezbollah had taken U.S. hostages. So Reagan hooked up with a charlatan named Ghorbanifar to be an intermediary with the Iranians, trading arms for hostages, even after he had emphatically declared publicly he would not deal with terrorists.[11] The Iran/Contra crisis was Reagan's worst nightmare, and he violated two laws in exercising what he thought was his executive privilege. In semi-repentance after the Tower Commission released its negative report, he admitted he didn't know all the details of what happened on his watch, an equally damning personal indictment of incompetence.

Unlike his penchant for the military, Reagan's aversion to domestic government was limited to reducing taxes, the opposite of the first course of deficit reduction, or reducing spending. If a household has its income cut, from loss of a job or other misfortune, and does not reduce its spending habits, increasing debt is the result. When Reagan came into office the top rate income tax was 70%. Reagan reduced corporate and higher income taxes to 50% and then in 1986 reduced it further to 28%. The top corporate tax rate fell from 46% to 34%.[11] The subsequent rhetoric was that this was pro-business, pro-family, pro-labor and pro-liberty. But it was also anti-deficit which nobody wanted to acknowledge.

Reagan received less than ten percent of the Black vote, primarily because it was clear he advocated a southern sympathy. In an effort to win Black voters his political team in 1981 wanted to educate Reagan about Black issues like affirmative action, voting rights, small business rights, housing, etc. Reagan had wanted a delay in the *Voting Rights Act* that Congress had sought for a ten-year extension. In correspondence from Elizabeth Dole to James Baker, the White House Chief of Staff, the Reagan White House inaugurated a Black strategy that focused on policies, liaison and affirmative action.

The problem was Reagan wasn't really that interested in promoting Black issues. But his staff convinced him of its political importance so he capitulated. One huge result was an Executive Order on September 15, 1981 that supported Black colleges. When the Office of Management and Budget cut education funding by about 40%, there was a scramble to remedy and restore the president's initiative for Black colleges.

Reagan's other domestic political leaning, his objection to government regulations, was revealed in his withholding approval of S.1128, "The Water Quality Act of 1986." Reagan's argument is that the proposal "exceeds acceptable levels of intended budgetary commitments," according to his November 11, 1986 disapproval. This argument was clearly a ruse and excuse for his antipathy for government regulations. He really did object to expanded federal enforcements even when they clearly meant improved public safety. His Task Force on Regulatory Relief made additional plans for derailing other federal regulations.[12]

"Franklin Roosevelt was the first president I ever voted for, the first to serve in my lifetime that I regarded as a hero," said Ronald Reagan in January 1989 when he was out of office.[13] As a young man he had seen Roosevelt when he visited Des Moines in 1936. Like the jaunty, ebullient, friendly and confident FDR, Reagan too enjoyed un-matched personal popularity that often transcended his forgetfulness, frequent incompetency and dementia.

I was too young to experience any of FDR's activities, except the ones about World War II. But my working class family—what I call the bib overalls class, or less than blue-collar workers—all liked him for his policies that gave people hope. And I liked Reagan for his personal optimism, but disliked him for his disastrous policies, both foreign (except for the treaties with Russia) and domestic.

Roosevelt and Reagan, immigrant sons of the American soil, one Dutch the other Irish, shared much more in common than the differences in their party affiliation might suppose. Both had been governors, both were practical politicians, one a Democrat, the other a Republican, one with a physical disability, the other with a mental disability, Alzheimer's disease.

Although representatives of the opposing liberal and conservative strains in American politics, they were both masters of practical government and compromised on ideological policies when it proved beneficial for the public. Roosevelt created a vision of how government could rebuild society after the collapse of the private sector and the economy beginning in 1929. Reagan always had a core belief about the dangers of government encroachment, but became a pragmatist when he saw that government responses had to be balanced with the private sector to sustain American society. Both were part idealist and part realist. They thought in foreign policy directives, each one loathing tyrannical and dictatorial states, whether Hitler's Germany or Russia's communist leaders, and the dangers posed when national leaders seek global domination. They are examples of the Alpha male American.

The Roosevelt Memorial In Washington DC.

The Reagan Grave at the Presidential Library (Simi Valley, California).

Chapter 10

WEATHERING CLIMATE CHANGE

Is there not the earth itself, its forests and waters, and all other natural riches, above and below the surface? These are the inheritance of the human race, and there must be regulations for the common enjoyment of it.

(John Stuart Mill, 1848, p. 161-62)

Earth has another threat, equally damaging and potentially longer lasting than the threat posed by nuclear annihilation during the Cold War, and also caused because of human actions. Inhabitants have over time brought about the slow erosion of the planet through neglect and carelessness. A combination of activities are converging that together form a dire scourge that could mean the slow eradication of organic species, including humans. The twin issues are population explosion and the effects of climate change on the environment.

There are 7.3 billion people on earth at this writing, nearly a three hundred percent increase of human primates in my lifetime, and the numbers are not declining. The earth's domination by just one species has not existed for 65 million years when dinosaurs roamed everywhere. Now population increases at current levels are unsustainable if humans hope to maintain earth as providing enough food, water and environmental natural resources to keep everyone healthy and viable. Moreover, the unremitting pace of consumption, waste and environmental change has stretched the planet's capacity that can only precipitate more environmental catastrophes, which already occur throughout the world.

Those who maintain that population pressures on the environment, and that energy corporations do not influence planetary climate changes, have their heads in the sand. Carbon dioxide, the principal chemical contributing to the rise in earth's temperature, creating disturbing and disruptive changes in nature such as hurricanes, tornadoes, violent storms, droughts, rising heat temperatures, retreating glaciers and rising sea levels, has reached new levels of alarm. In August 2015 the temperature in the Pacific Ocean near the equatorial coast of Peru reached 3.42 F., warmer than the historic average, and the highest ever recorded.

Weather is temporary, constantly changing. Climate is a long-term scientific term. Climate is the thermostat of the planet, varying over time, particularly when the oceans become warmer, as they are now. Oceans warm and glaciers melt because of the global increase in carbon dioxide, or CO_2, the so-called greenhouse gas. Without carbon dioxide there would be no life on earth as we know it, and the land would be frozen, as it is on other planets. With too much of it, the earth becomes increasingly hotter and climate and weather changes occur.[1]

Carbon in the atmosphere, that causes global warming, passed four hundred parts per million in March 2015, the highest level in the earth's history. Natural variability alone does not explain these dramatic temperature changes. Over the past 650,000 years, the earth never had more than 300 parts carbon dioxide per million. It has now reached more than 400+ parts per million. The hottest years on record have been the past few years, with each year setting new records.

The Kyoto Protocol is a now dormant international treaty, extending a UN treaty from 1992, to reduce the so-called greenhouse gas emissions, an acknowledgement that global warming exists and that all states need to do something to reduce it. While 192 parties did sign the treaty, the U.S. was one of two nations that failed to re-sign. A further treaty, known as the Doha Amendment, witnessed by 37 nations created binding target dates for carbon reduction emissions. In 2014 additional negotiations were held in Lima, Peru to agree on further agreements beyond Kyoto. But the U.S., China and India, the greatest contributors to carbon emissions, would not ratify.

Consequently the most anyone can expect in the near future, as even Pope Francis noted in his 2015 encyclical *Laudato Si'*, is superficial rhetoric, sporadic acts of philanthropy and perfunctory expressions of concern. Any genuine attempt to introduce change is viewed by many conservatives in the U.S as a nuisance based on romantic delusions, or an obstacle to be circumvented.

We have already passed the era of consequences for what we have done to endanger the planet. The American Petroleum Institute, the lobbying group for energy companies of oil and natural gas, has convinced legislators with campaign contributions and the public with propaganda that there is doubt about the scientific facts. As a result, Congress has stalled and has failed to acknowledge results by world scientists. And yet politicians, because of campaign fund contributions to their re-elections, and the bombardment of propaganda from energy corporations, have cravenly chosen to follow the moneyed instinct and not prioritize the safety of its citizens.

Pharmaceutical companies always ask us in TV ads for their products to "ask your doctor" about taking certain prescribed drugs, with the implicit assumption that we will abide by that advice. There is no corresponding "ask your scientist" when comes to getting scientific consensus agreed to by the public, if the issue, like climate change, conflicts with prejudices and propaganda from energy companies averse to science when results threaten profits.

A part of this ingrained bias against selective scientific issues, like evolution, is one consequence of the so-called democratization of the Internet where one can always find some source that appears to be authentic to support nearly any view. The former societal view was that there were certain gatekeepers of knowledge, like teachers or professors, who knew the facts and dispensed them in formal settings. Now if it is written somewhere, like on a blog, or visible, like a YouTube video, anyone can find a source that will confirm any bias, yet the information may not be validated science, but just an unconfirmed opinion.

Let this anomaly represent the challenges American society faces as it is more skeptical of validated scientific knowledge, and hesitates to accept the belief of the consensus of the scientific community. There is a wide gap now between known scientific evidence and beliefs that run contrary to them. Climate change is just one of them.

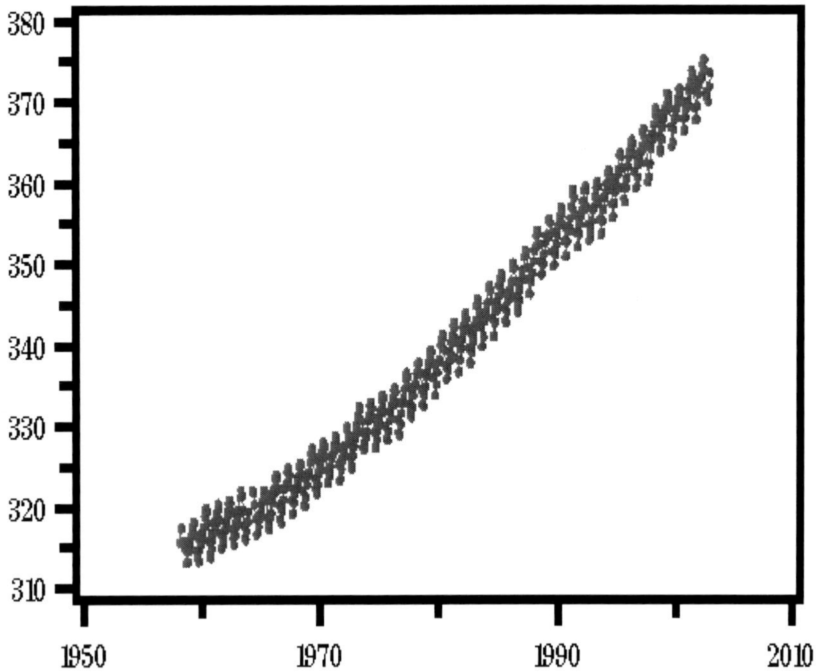

Scripps Institute Measurments of Co_2 at Mauna Loa, Hawaii as of 2010.

For example, the story of the Aral Sea, millions of years old and once the fifth largest lake in the world, is one of the worst historical cases of a man-made environmental damage. The Aral Sea, used for centuries by South Asian peoples as a water source, was filled at the south end by the Amu Darya, and at the north end by the Syr Darya. Beginning with the collective farming policies in the 1930s, the USSR began diverting water from the Aral Sea into Kazakhstan and Uzbekistan for producing cotton which requires large amounts of water to grow, but whose product brought in foreign revenues.[2]

Over a generation the water in the Aral Sea became heavily salty and increasingly empty as more water was extracted than was flowing in for replenishment. Fish and the fisheries industry died. The sea is all but dead today and efforts at its resurrection are too futile and too late. Nearby residents now grapple with health problems from the pollutants blowing in the wind from the detritus at the desiccated and exposed seabed. The Aral Sea is now a prime example of how human activities, overconsumption and diversion, can significantly disrupt a large natural water resource.

The Aral Sea for millennia was once one of the world's largest inland bodies of water—at 26,000 square miles, slightly larger than the state of West Virginia—and its manmade destruction is now complete.

There are other examples around the world. Lake Mead is at its lowest capacity ever as the Colorado River reaches new low levels, and as more people and greater use of water for agriculture sucks its once plentiful supply. The city of Las Vegas gets 90% of its water supply from Lake Mead. The last year Lake Mead was full was in 1983. The long drought has reduced Hoover Dam's hydroelectric energy output.

In 1993 Lake Chad in central Africa was one of the largest lakes in the world. Today it is non-existent.

On the other hand, judicious use of water resources can be productive. The Israel Water Authority sets quotas for each farmer and taxes excess use at a much higher rate. Desalination plants in Israel have increased useable water in the past years, and they now produce 40% of the country's water supply. Additionally, Israel purifies 85% of all household wastewater. Drip irrigation for crops began in the 1960s.

There are over 15,000 desalination plants worldwide in 150 countries, and at this writing only one in drought-stricken California (in Carlsbad, and another one under construction in Huntington Beach) providing 80 billion gallons of drinking water daily to 70 million people.

The desalination plant in Carlsbad, California, the largest such plant in the Western Hemisphere, converts 56 million gallons of water daily into drinking water for over three million San Diego County residents. It took 15 years and one billion dollars to build, and will be paid for in increased water taxes, but it portends the future. It operates with 16,000 reverse osmosis membranes, essentially salt filters, as water is forced through high-pressure pumps through perforated pipes buried in the sea floor. Fish screens keep out all but the smallest marine organisms. Before passing through the reverse osmosis filter, the water passes through a sand and gravel filter, then a charcoal filter and finally a fine filter screen.

The problem with desalination plants is that they have to literally suck in almost twice as much water as they produce. The plants have to return the discharged water back to the ocean with heavier salinity that was extracted.

Like Israel, all states and localities need to unify fractured and divisive local water authorities and unify them under a state water authority so there are comprehensive strategies for usage and conservation. The water projects include storm water capture, new storage capabilities and groundwater treatment. Certainly, there has to be enough water to quench the thirst of people, and produce the food that feeds them.

Energy: The Engine That Connects Us

How much energy do we Americans use on average every day? According to Professor Sally Benson, Professor of Energy Engineering at Stanford University, and former Deputy Director of the Lawrence Livermore Berkeley National Laboratory, we use about 250 kilowatt hours daily, the same amount that would keep a hundred 100-watt light bulbs lit for 24 hours. That may not sound like much, but energy permeates our lives and not just in household lighting but in cooking, refrigeration, cleaning, computing, commuting, among other common uses like transportation, in buildings for heating and cooling, and in a variety of everyday machines. Our lives would be more primitive without energy. And unless we live on enough land and have enough resourcefulness and creativity, energy of this livable magnitude is not something we can produce by ourselves.

The problem is that we are using excessive fossil fuels, particularly cheap coal, to generate energy, emitting too much carbon into the atmosphere. The planet is warming and the long-term effects are grim. According to the U.S. Energy Information Administration, coal contributed 37% of all U.S. electricity in 2012, 30% came from natural gas, 19% from nuclear power plants, 7% from hydropower sources, and 5% from renewable sources such as wind and solar. Those percentages need to be reversed with a majority of energy emanating from renewable sources.

On the other hand, we will always need some petroleum products to run the cars, trucks and airplanes that cannot operate on other energy sources.

We cannot see climate change visually, cannot see the hole widening in the ozone layer, cannot personally gauge the *El Nino* effects in the Pacific where carbon seeping into the seas and warming the water causes more troublesome storms, or witness the melting of the glaciers elevating ocean levels which threaten coastal cities, or see the loss of animal habitats. One informed estimate is that the seas may rise as high as 10 feet in the next 50 years, water that will swamp most coastal cities. Repairs with seawalls and other protective mechanisms, like those today around New Orleans, will be ineffective because the water will continue to rise. The Arctic ice cap diminished by 40% just in the years between 1970 and 2005, and the West Antarctic ice sheet is melting as quickly.

If you live in a mega-metropolis, you can also breath in insalubrious and toxic air regularly from industrial and vehicle emissions.

Climate scientists are warning us of the planet's rapid and destructive alterations in temperatures and in the accumulation of greenhouse gases that will lead to the decimation of insect and animal species and, most significantly, will damage global food sources. The primary scientific evidence for climate change is embodied in a variety of environmental studies. Rising temperatures are the most closely monitored. The temperature has risen 1.03 degree F above the average in the 20th century. An overall rise of global temperature by two degrees is irreversible and will cause untold damage.

Heat and acidity have damaged the coral reefs that have declined by 40% since the 1970s, greatly impacting marine life and thus the marine food chain. According to the *National Geographic* in June 2015, as many as 25% of marine fish species live in coral reefs, and one-eighth of the world's population relies on fish from coral reefs to sustain a food source and income. Half of the world's coral reefs, composed of part animal and part mineral deposits, have already died. A coral the size of a large boulder could be 500-1,000 years ago and be home to a variety of marine organisms. Charles Darwin was one of the first natural scientists to analyze them in his 1842 book, *On the Structure and Distribution of Coral Reefs*.

Some fish are fleeing their normal habitats because of oceanic heated waters. Bottom-trawling nets in search of fish have scraped the seafloors disturbing a variety of marine life. Moreover, seabed mining has scarred ocean floors and chased away marine life. All of these un-noticeable-to-the-naked-eye conditions are definitely manmade, creating environmental and fish habitat damage that disrupts a major food source.

Poachers in Africa roam into gorilla wildernesses seeking animals to kill and sell on the black market for exotic foods. Trees are cleared for planting crops, and for firewood for cooking diminishing the land available for livable habitat. Because of their loss of living space, chimpanzees are emerging and having more interactions with humans, to their disadvantage, as many raid farms for food and are killed by avenging farmers. All non-human primate populations everywhere are endangered.

Because of our skewed, peculiar and largely dysfunctional politics, radical conservatives as a political group, have not only not agreed to this scientific consensus, they have worked to deny, remain silent about, and actively work against the negative effects creating climate changes. Elected officials in many southern states—for example in Kentucky, Louisiana, Texas, and Oklahoma—are the tools and faces of the energy companies of oil, gas and coal.

As President Obama was preparing sweeping rule changes sponsored by the EPA to mitigate climate change, five Republican governors, two of whom were presidential aspirants, were refusing to carry out the federal rules in their states. The Senate majority leader, Mitch McConnell, wrote letters to all governors to refuse to carry out the rule changes.

Yet nothing exemplifies widespread scientific opinions and the overweening profit motives of special interest groups as the controversy about climate change. Eco-activists, environmental earth huggers, doomsday apocalyptic prophets, petroleum lobbyists, population theorists, believers in divine plans for earth—all have separate views about what climate change is and how it affects the future of the planet. As Pope Francis noted in his 2015 encyclical:

> *But a sober look at our world shows that the degree of human intervention, often in the service of business interests and consumerism, is actually making our earth less rich and beautiful, ever more limited and grey, even as technological advances and consumer goods continue to abound limitlessly. We seem to think that we can substitute an irreplaceable and irretrievable beauty with something we have created ourselves.*[3]

There are even more distressing and irrational examples. Professor Asad Zaman, Vice Chancellor of the Pakistan Institute of Development Economics, former graduate student and lecturer at prestigious American universities, gave an closing address at a sustainability conference sponsored by the economics department of Fatima Jinnah Women's University in Rawalpindi, Pakistan on April 4th, 2015, at which I was a sponsored attendee. He delivered an alternative to the standard economic model. He deplored the secular society that favors individual freedom, criticized the capitalist model, which he claims has destroyed the planet, regaled against greed, deplored "the loss of brotherhood," and as an aside stated that American democracy was a sham. He used the worse examples of popular movies to be revealing of social and cultural actual facts.

His rambling speech was less of an economic presentation than an advocacy for Islamic values about climate change. That such religious penetration of one of the world's most significant problems, climate change, can occur at a conference intended to enlighten the audience with scientific data and conclusions reveals the depth of how much Islam has infused the most quantifiable of scientific analysis, economics, in an Islamic democracy. The fact that he took no questions and no one criticized his approach spoke volumes about the fear and intimidation of religious authorities, even at a scientific conference.

Yet as of 2015 more than half of Americans, and nearly 90% of scientists, agreed that global warming was caused by human behavior. But only a small percentage of Republicans in Congress have been willing to be on the record to accept that consensus. They repeat the banal mantra that they are not scientists, a code phrase that means they don't want to discuss the issue, because then they would be asked what they plan to do about it. Dealing with it legislatively would mean severely reducing carbon emissions, and that would cost the fossil fuel company providers dearly, a testimony to the power of the campaign contributions of those industries.

According to a Gallup Poll in early 2014, most Americans, 65%, believe global warming is currently happening or will happen in their lifetimes. Far fewer, 36%, predict it will be a serious threat to their way of life. A poll conducted by *The New York Times*, Stanford University and Resources for the Future reported in 2015 that the majority of voters support government action to curtail global warming. It also found that two-thirds are likely to vote for candidates who support fighting climate change. That includes 90% of Democrats, 78% of Independents, but only 48% of Republicans, most citing the possibility it will hurt the economy.

Five Republican presidential candidates in the 2016 election were a part of the science denial group whose studied silence on the topic or outright denial we have to interpret as possibly literary ignorance, if not outright hostility to science. Americans don't seem to be outraged by this kind of infuriatingly, publicly-stated, posturing ignorance of elected officials.

What a profound indictment of sensible dialogue—and congressional halls are supposed to be the venues of debate—is this kind of empty, negative, valueless formula by conservative American leaders for the greatest peril facing the earth and everyone on it. Once Republicans became a majority in the House and the Senate in 2015 they immediately declared they would subvert regulations aimed at curbing carbon pollution. In the past the coal industry, which supplies 90% of Kentucky's electricity for example, has had a stranglehold on regional politics and is resistant to regulations on carbon emissions.

A *New York Times* investigation in 2014 uncovered energy lobbyists making direct campaign contributions to state attorney generals who then use state power to fight federal regulations protecting the environment.[4] Republican state attorney generals in 26 states in 2015 worked to derail major federal environmental actions: to curb fish kills, to reduce ozone depletion, to eliminate the reduction of coal ash, and in general to loosen the efforts at reducing climate change policies.

In a democracy, the voice of the energy companies is as important as anyone else's voice. But a case can be made that undue influence of state employees and civil services, publically advocating for the interests of private industry that might also harm the public's safety over federal environment regulations, appears a bit tawdry. The shameless use of public offices to further private industry with substantial financial interests in defeating federal actions is unprecedented and borders on culpability of office.

A significant majority of Americans see economics as one of the most important issues, but agree that unless we take care of the environment there won't be much of an economy to enjoy.[5] In most polls they relegate climate change near the bottom of priorities despite the clear and scientifically uncontested evidence.

Who we are as primates also includes our temporary residency on a planet now under siege because of misguided beliefs and government policies that encouraged corporate development, and have managed less well the personal health and safety of the globe's inhabitants.

We are inhabitants of the earth but find that we are powerless to alter the perilous course of managing the earth's decline. We can see the origins of our problems in coal-fired power plans and the emissions coming from smokestacks. So we are not unaware of where the atmospheric damage comes from. Yet we need our electric hair dryers, microwaves, entertainment platforms, and lights at night.

I am pessimistic that we can regulate the continued polluted contamination impacting on our planet that is inevitably leading to a major wound on the earth bringing impaired living conditions to millions. We are like Sisyphus pushing the boulder up the mountain only to have it roll back down because we didn't finish the task.

Science Evidence

E.O. Hulbert published "The Temperature of the Lower Atmosphere of the Earth" in the journal *Physical Review* in November 1931. Here is what he noted then:

> *Calculation shows that doubling or tripling the amount of the carbon dioxide of the atmosphere increases the average sea level temperature by about 4° and 7°K, respectively; halving or reducing to zero the carbon dioxide decreases the temperature by similar amounts. Such changes in temperature are about the same as those which occur when the earth passes from an ice age to a warm age, or vice versa. Thus the calculation indicates that the carbon dioxide theory of the ice ages, originally proposed by Tyndall, is a possible theory.*[6]

Decades ago scientists understood the inherent dangers of pouring too much carbon into the atmosphere. Scientists in 1982 began raising alarms about rising earth temperatures. *The Houston Post* on March 3, 1982 headlined this story, "Rising Temperatures on Earth Could Threaten Human Life Congress Told."

A significant number of scientific reports—New Climate Economy Project, the Intergovernmental Panel on Climate Change, International Monetary Fund papers, reports in journals like *Nature Geoscience*—and newspaper accounts have succeeded in partially instructing the public about the consequential dangers of repeated release of carbon products into the atmosphere. There has been an exponential global rise in carbon dioxide emissions.[7] According to the 2014 dire report of the Intergovernmental Panel on Climate Change:

> *Human influence on the climate system is clear, and recent anthropogenic emissions of greenhouse gases are the highest in history. Recent climate changes have had widespread impacts on human and natural systems . . . Warming of the climate system is unequivocal, and since the 1950s, many of the observed changes are unprecedented over decades to millennia. The atmosphere and oceans have warmed the amounts of snow and ice have diminished, and sea level has risen.*[8]

We may experience changed weather conditions where we live, and have probably read about how the weather is not normal anywhere. But we may have a hard time understanding the exact cause and effect relationship between the altered weather and human effects on the planet's ecosystem that modifies the weather. The stark reality is that since 1950 the globally averaged combined land and ocean surface temperature has more than doubled. Global carbon emissions have more than quadrupled.

I turn again to the scientific consensus of the Intergovernmental Panel on Climate Change and its 2014 report:

> *Anthropogenic greenhouse gas emissions have increased since the pre-industrial era, driven largely by economic and population growth, and are now higher than ever. This has led to atmospheric concentrations of carbon dioxide, methane and nitrous oxide that are unprecedented in at least the last 800,000 years. Their effects, together with those of other anthropogenic drivers, have been detected throughout the climate system and are extremely likely to have been the dominant cause of the observed warming since the mid-20th century.*[9]

Some of the remedies for mitigating the ineluctable advance of the effects of severe climate change are clear and the risks to humans, animals and plant life are enormous, yet special economic interests resist them. Other solutions are not so obvious. The technology is already established to switch to non-fossil fuels. But technology itself might not be enough to reverse the trend.

Yet partisan backers and advocates of the interests of major corporations block any legislation that causes the U.S. to act in unison with other developed nations to reduce carbon emissions. Both partisan politicians and corporate executives defend their unjust causes with unsound science and passion. Even China is advancing solar and wind power alternatives to reduce carbon outputs.

The petroleum industry's protection of its self-interests is the principal obstacle, all misdirected propaganda and risible claims to the contrary. Slowly but inevitably, the public is coming to the table of truth about the planet's atmosphere and the air we breath, and that moves the wind currents and warms the ocean. As of 2014, 74 nations and more than a 1,000 investors and businesses had signed on to a declaration by the World Bank calling for an international price on carbon. More than 40 countries have already placed a price on carbon. The U.S. is not one of the signatories, though nine U.S. states are.

Nevertheless, in November 2014 the U.S. and China signed an historic agreement. America set a new target of cutting net greenhouse gas emissions 26% to 28% below 2005 levels by 2025. President Xi announced plans to increase the share of renewable energy and nuclear power that China uses to roughly 20% by 2030. Although the U.S. is the largest contributor of greenhouse gases, China and India combined with soon surpass the U.S. and Europe, so cooperation is essential.

The Intergovernmental Panel on Climate Change I referred to above says that greenhouse gases "will cause further warming and long-lasting changes in all components of the climate system, increasing the likelihood of severe, pervasive and irreversible impacts for people and ecosystems." In effect this means that species will become extinct, political systems will remain petrified and uncertain, and people will die avoidable deaths. As is true in all disaster scenarios, the poorest on the planet will feel the effects the most and the earliest. According to a World Health Organization report in August 2014 (No. 266), between 2030 and 2050 climate change is expected to cause 250,000 additional deaths per year from malnutrition, malaria, diarrhea, and heat stress.

Fossils as Fuel: Burn, Baby, Burn

President Obama in August 2015 proposed new regulations for the Environmental Protection Agency (EPA) that required states to reduce greenhouse gas emissions from power plants by 32%. Power plants are responsible for a third of all carbon emissions, more than all of America's cars, planes and homes combined.

Obama's proposal, known as the *Clean Power Plan*, also called for states to draw 28% of all power from renewable sources. The Senate majority leader, Mitch McConnell, Republican from Kentucky, a major coal producing state, sent a letter to all governors, as I noted above, asking them to reject this proposal by refusing to comply. If climate change is the most important issue facing the long term health of the earth, why would he oppose steps that would reduce carbon emissions? The short answer—apart from bigoted opposition to a sitting Black president—is that politicians beholden to moneyed interests trump activities that further the "general welfare," that phrase from the *U.S. Constitution* that gets short shrift in radicalized politics.

Coal, that black stuff that coal mines gather from the bowels of the earth, is ancient vegetable material—dead trees, leaves, twigs, bark, that all sank into a one-time swamp, decayed and formed peat that cooked into coal, a seemingly endless amount of cheap fossil fuel. It has always seemed ironic to me that we need to essentially burn the world's dead vegetative past, our fossil fuels, in order to literally consume our considerable energy needs. What strikes me as ironically probable is that we, the sum total of all human primates, may well be a part of that same kind of past—fossils that some day become the energy source for subsequent organisms need to survive. We awaken the dead past, so to speak, that in excess in a different state may lead to our own deaths.[10]

Similarly like coal, oil is formed from ocean plankton, the sea's tiny organisms, one of whales' favorite food sources. Sea-beds turned into oil because of millions of tiny organisms and became fossilized fuel. Because these ancient life forms contain carbon, the release of carbon dioxide—one part carbon and two parts oxygen, odorless and colorless—from the burning of fossil fuels is the chief component now warming the earth. Carbon is also integral to human primates and a part of all living organisms. If you extract the water from human bodies, the remaining dead weight is mostly carbon.[11]

Protecting the earth against the ravages and revenge of the natural world and the extinction of untold numbers of species like walruses, snails, bumblebees, deer, is important. It is not as important as the deterioration of human food sources, mainly wheat, rice and corn. According to the Intergovernmental Panel on Climate Change, average food yields are declining and average yields in the future are less certain.[12]

Developed countries don't want to reduce their economies to confront this rush to planetary disaster. Developing countries want to be able to build their lesser economies to achieve the kind of lifestyle available in developed nations. Partisanship and corporate lobbying has ensnared domestic U.S. politics making legislative gridlock its frequent accomplishment.

We can all observe or believe implicitly that the glaciers are melting, and that the weather everywhere is inconsistent—hotter, colder or stormier depending on where one lives. The summer months of 2014 were the hottest on record . . . until the summer months of 2015. It's another step to infer what these more unstable weather changes, and other impending calamities like carbon emissions and the accompanying rising temperatures actually mean for the planet. A trend analysis for the whole of the earth, and how it affects all humanity, like food production and a sufficient supply of water, is not easily obtained. But we do know that greenhouse gases will double in the atmosphere by the end of the century, pushing up average temperatures everywhere by 3-8 degrees.

Unlike the human primate, nature is not abrupt in its modifications. Nature's moves are gradual, incremental, slow to show long-term intent. Recall how long it took for the evolutionary process to achieve the state of development a human primate is. But glacial really does describes the political will to arrive at any universal solution.

Opposition to Government Environmental Regulations

Let's explore some of the reasons why this oppositional, anti-government intrusion position has been an integral part of radical political strategy for decades ever since President Reagan.

The Reagan governing philosophy, that became the typical conservative political posture, was to reduce or eliminate government regulations that interfered with business and industry, and to give them freedom to perform, to promote economic expansion, and to turn federal actions over to the states and localities. Reagan's arguments invariably were that government was spending too much money on programs that regulated businesses.

President Nixon established of the Environmental Protection Agency and its laws protecting the environment with the *Clean Air Act* and the *Clean Water Act*, both epic changes in domestic policy. A Republican president established the EPA, but a subsequent Republic president, Reagan, perpetually opposed legislation that enhanced the quality of the public's safety under EPA by favoring businesses and polluting industries that continually threatened the environment with increased carbon emissions. That conservative strategy did not change over the succeeding decades.

Under the Reagan Administration, opposition to all government regulations intensified. The *Clean Air Act* came up for congressional renewal in the summer of 1981. Congress proposed to reduce high sulphur and coal producing carbon emissions, and vehicle emissions like carbon monoxide and nitrogen oxide. The auto industry sought to turn back existing regulations and all future ones EPA rules. At the time, 81% of the American public agreed with changes in air quality.

Reagan, however, siding with his unwavering anti-federal beliefs, and regardless of what the general public wanted, threatened to veto the passage of the bill because, as he claimed, the law "diminishes the role of states in determining their own air quality programs," and the standards are "too stringent, based on poor science, protect against insignificant health effects, and wasting billions of dollars."[13] The White House strategy was to let the private sector carry the "political load."

Some of the businesses who sought continual relief from all EPA regulations under the Reagan Administration at the time were Conoco, AMC, US Steel, Monsanto, Ford, Procter & Gamble, GM, CMA, among others. These were the companies the public least trusted, according to contemporary reports.[14]

Industries not only seek to get relief from government regulations but purposefully pollute to escape costly procedures to eradicate toxic waste. Tesoro is a petroleum company based in San Antonio, Texas but has refineries throughout Canada and the U.S. It controls supplies gas stations under the name of Tesoro, Arco, Exxon, Shell, USA gasoline and Mobil.

For seven years the owner of the Tesoro refinery in Martinez, California unlawfully allowed butane, propane and other gas emission fumes to enter the atmosphere. It failed to run the fumes through filters as required and let the fumes drain through the refinery's sewer system and around the installed filter system, imperiling both the workers and the planet. Tesoro was fined a paltry $4 million for this infraction. Without inspectors, petroleum corporations and industries would, and do, skirt environmental laws putting everyone's health in jeopardy.[15]

By 1983 a number of scientists concluded that there was a direct link between air-borne industrial pollutants, acid rain, and the destruction of fresh water life. Nevertheless, ignoring results from the scientific community, the Reagan Administration opposed congressional efforts to reduce sulphur emissions by 50% because of industry expenditures. One White House internal memo noted about the news of the increase in carbon footprints was benign neglect, "Let it lie, in hopes it is already dying."[16]

In 1986 the *Clean Water Act* came up for re-authorization in Congress. It passed the House by a vote of 408-0 and the Senate by 98-0. The people had spoken unanimously. But Reagan, not deviating from his misguided principle, threatened a veto. Several governors wrote to Reagan urging him not to veto, including governors Bill Clinton, and Republican governors John Sununu, Lamar Alexander and Thomas Kean. Reagan realized he could not survive a veto over-ride so he signed the bill reluctantly.[17]

Reducing regulations on public safety was equally imposed on coal mine operations. In 1980 coal mine deaths totaled 133. In 1981 the deaths topped 153. The increase in mine safety apparently had no effect on the Reagan Administration. The Reagan Administration's Office of Management and Budget produced cutbacks in mine safety so severe that even Elizabeth Dole objected. Worse, there were personnel cuts in mine safety citations, and mining companies laid less in fines. Inspectors were personally threatened while in the field and during inspections.[18]

True to his ideology and his obsession against government interference, Reagan's deregulation policies extended across all federal government agencies and diminished the public welfare. His typical reasoning was that deregulation "streamlined" government operations, emphasizing a positive note instead of the actual negative effects on public safety. There were more deregulatory edicts.

On January 12, 1983, the Department of Transportation issued a delay on constraints of heavy-duty trucks, and amended the rules on the transportation of hazardous materials.[19]

The EPA streamlined its registration process by reducing the burden on industries by allowing it to issue no statement or evidence about whether or not public safety was affected.

The Bureau of Land Management on May 23, 1983 reduced the regulations on those applying for oil and gas explorations.

The Federal Highway Administration permitted states to increase the time for inspections of bridges.

The Food and Safety Inspection Service allowed a wider variety of chemical agents to be approved for processing food product containers.

Finally, the Board of Governors of the Federal Reserve, under pressure from Reagan, reputedly streamlined the regulations by extending securities credit to banks on August 3, 1983.

These conservative policies of deregulation favored the moneyed interests at the expense of public safety and the general welfare, especially now in time of planetary climate stress, corporate intransigence, and rampant anti-government sentiments, particularly among the political admirers of the Gipper's obsession with a misguided and undeviating ideology. Mother Earth will be indifferent to futile political opposition to Obama's *Clean Power Plan* or any plan that attempts to control for climate change, and unkind in her retaliation.

WATER SHORTAGES: THE CASE FOR CALIFORNIA

There will always be controversy over the cause of drought and water shortages because so many political leaders do not believe in the science of climate change. However, what is clear is that there is increased need for water for agriculture, and water to service a larger population. Unquestionably, temperature changes have decreased snowpack and worsened evaporation from reservoirs. California's continuing drought, the latest water shortage in one of many western states that will experience more of it in the near and far future, became more apparent in 2014.[20]

Snowpack provides about one-third of California's water for farms and cities. Climate change is depositing less and less snowpack in the Sierra Mountains, bad news for water users and farmers who relied on its once plentiful source. Yet the state had only 12-15% of its normal water average in 2014. And 2013 was already the driest year in California's history.[21] Shasta Lake waters were 65% below its historic average. The water for wildlife refuges in central valley in California was down 30%.[22]

The average California household uses 362 gallons of water daily, and residents are slow and reluctant to diminish water use, even for the lawns of Fresno. And water is cheap and its price is usually fixed for the delivery and maintenance of the pumps and pipes. Water will obviously have to be rationed, as food and gas were in World War II, but it may also have to be taxed for its use so the public will realize it is a precious commodity that should be rationed. Governor Jerry Brown finally cut the use of water for residential use by 25% in May 2015.

California's agriculture produces nearly half of all the nation's fruits, vegetables and nuts, half of the nation's milk and cream supply, and half of the grapes. And for this food largesse agriculture consumes 80% of the water. California agriculture is the consumer food heartbeat of the nation, a state with 81,700 farms with an average size of 311 acres, and a total spread of 155,959 square miles.

Alfalfa is one of California's products that suck up a disproportionate share of water. Its primary use is for cattle feed. The water footprint starts with vegetation, moves to animal feed, to animal slaughter, and then on to processed meat and byproducts. In this sequence, 145,000 gallons of water from lakes, aquifers and rivers is needed to produce one ton of beef.

Decidedly, increased population will demand more water than now exists for the California's existing population of 38 million, estimated to grow to 46 million in 2035. The present drought will only worsen the condition of a larger population and the further need for the agriculture the state produces.

Cities, particularly smaller rural California cities, have run out of water. Big cities like Los Angeles with close to 10 million people do use lots of water. Surprisingly, however, many cities like Sacramento do not have water meters so officials don't know how much water individuals are actually using. Nobody seems to know how much the aquifers are being depleted because few have meters that measure it.[23]

Fields, once high with native grasses, are brown so cattle cannot feed.[24] The lakes ran dry years ago. Now the rivers and streams have dried too, and the aquifers are becoming depleted. It may be time to bring back outhouses to save on flushing toilets.

All up and down Highway 5 that runs north and south through the agricultural heartland, signs, like this one, "Congress created the dust bowl," blame politicians for the water shortage. No sign mentions the lack of rainfall in the mountains because of climate change, or overuse or conservation by farms or urban residents, or the hard luck of rain dances. Farmers find it more politically expedient to blame politicians, reputedly for advancing too far the cause of endangered species at the expense of agriculture.[25]

A combination of population increases, agricultural imperatives and privileges, a shortage of snow and rain in the high mountains, and lack of water conservation measures have all contributed to this emergency. As soon as a few wells and water silos run dry and no rain falls, everyone will panic and begin serious water conservation procedures. As John Wesley Powell was keen to point out in his explorations and writings in the 19th century, the West has always been a desert. No one, certainly not politicians nor their constituents, seem to take adequate precautions until it is too late.

California has what most western states lack—seawater. Like Persian Gulf states that have water only from desalinization plants, California needs to begin the development of more such facilities along the coast. At this writing one such facility is operational, and one scheduled to be reopened. The United Arab Emirates produces 564 million gallons of water daily from the sea. It's time to think way outside the water towers, water coolers and depleted aquifers, and to think like desert people.

But it isn't just about California. The Colorado River begins in the Rocky Mountains of northern Colorado and winds through the Grand Canyon. It supplies water to the whole of the Southwest and the states of Colorado, Wyoming, California, Nevada, Arizona, and New Mexico, yielding its precious resource to 40 million inhabitants and four million acres of farmland. Recently, the Colorado River has shrunk to little more than a trickle.

I watched its slow movement at various locations along Highway 95 in 2015 where it is the border of Arizona and California, its once mighty flow reduced from its former days. By the time Colorado River water reaches the Palo Verde Diversion Dam about 10 miles north of Blythe, California, it is a shadow of its former rampaging self that has navigated through the Grand Canyon. The dam diverts the precious waters into canals for the rich agricultural lands along the wide plains just west of the Arizona border. This part of California uses Colorado River water provided by the Bureau of Reclamation.

After more than a millennium, the river is exhausted from more than a decade of drought, lack of mountain snowfall, increased summertime temperatures, and excessive overuse. Water rationing began too late. The Southwest, a vast desert, will have to set priorities and decide whether its urban inhabitants in Phoenix, Las Vegas, San Diego and Los Angeles are more important than its farming communities that need water for food production. Food will certainly become more expensive, a trend already noticeable.

As desert conditions not only prevail but increase everywhere, intellectual honesty and leadership requires the kind of strength and resolve epic heroes had. As it is, an intellectual desert abounds. Mere human decision-making often may not be enough. The devout may believe prayers alone will be enough to produce miracles. The more idealistic may believe that semi-divine and superhuman beings, like a Hercules, a Batman, Spiderman, Wonder Woman or Superman, were created to make us think that one hero can resolve our common problems. Realists know that a collective, unified, sacrificial action by everyone, as occurred in World War II among Americans, not just the leadership of one person, is the only successful method for combatting global environmental and social issues.

But there is a variety of evidence, besides glacial erosion and water shortages that reveals how delicate climate change effects organic life, for example birds. Of the 588 North

American bird species the National Audubon Society studied in late 2014, more than half are in danger of range extinction. Models indicated that 314 species would lose more than 50 percent of their current climatic range by 2080.

The report noted: "Of the 314 species at risk from global warming, 126 of them are classified as climate endangered. These birds are projected to lose more than 50 percent of their current range by 2050. The other 188 species are classified as climate threatened and expected to lose more than 50 percent of their current range by 2080 if global warming continues at its current pace."[26]

The study covered Alaska, Canada, and the lower 48 states. Dozens of North American birds winter in Latin America, but because researchers lacked sufficient data to study the effects of global warming on those birds' winter ranges, it is likely that even more birds are threatened by global warming than we have been able to identify so far.

The World Wildlife Fund in Mexico discovered that fewer monarch butterflies are migrating from the northern U.S. to Mexico. In 2004 an estimated 550 million butterflies completed the winter migration. In 2013 only 33 million made the journey. The biggest culprit is large-scale use of herbicides that destroyed milkweed, one of the butterflies' major foods.[27]

SUN AND WIND OPTIONS

The special theory of relativity has led to the conclusion that inert mass is nothing more or less than energy.

(Albert Einstein, 1916)[28]

Utility companies are being challenged everywhere as new technologies discover long-running business plans for how power is sold to customers. Even less developed countries are finding new ways of renewable energy production, to cut the health costs of pollution and use more efficient methods.[29]

Among the developed countries, Germany has established new renewable energy projects that have cut into the traditional non-competitive utility producers. Solar panels have been decreasing in price over the years and countries like Germany are buying into renewables heavily. As it discontinues its nuclear power sources, Germany seeks to have 30% of its energy coming from solar and wind sources.

Already, a third of all Denmark's energy comes from wind power. It has columns of turbines out in the Baltic Sea waters around it have been generating energy for businesses and homes. And Danes have been using this source for years. Venture capitalists of modest means form partnerships to buy into a single wind turbine whose owners sell the energy generated to the utility company for total grid use. Everyone benefits. My Danish colleagues confirm that Denmark will be free of carbon-generating fuels by 2020, an extraordinary feat.

Holland has installed solar panels in road surfaces that generate electricity. The experiment has exceeded expectations. Along a 70-metre track for bikes near the Dutch town of Kommerie outside of Amsterdam panels beneath the track generated more than 3,000 kwh, enough to provide a single-person household enough energy for a year.

Even some American states, impatient with the lack of congressional action on energy, have begun to set aggressive plans on their own initiative. A new outlook is forming globally as more and more individuals and governments grapple with the need for new sources of energy to move away from the tight grip of utility companies. New plans are in the wind, so to speak.

Developers operating in several western states automatically install solar panels on the roofs of new homes, retain ownership of the panels for two decades, but sell homeowners the power from their panels at 20% discount from the utility companies. Normally, utility companies charge higher prices during peak hours of the day when usage is greatest. Solar power does not rely on peak times—its greatest source of power is the daylight hours—and is therefore much cheaper.

Quinault Tribal Fishing Community

One example of a Native American community that is affected by climate change, but has not contributed to it, is the Quinault Indian tribe along the western coast of Washington State. The Quinault Indian Reservation comprises 208,000 acres, or about 300 square miles. To the immediate east are the lush forests of the Olympic National Forest. Along the coast are vast stretches of sandy beaches and rocky cliffs. Rivers from the mountain drainage plunge into the sea every few miles.[30]

Based on interviews with Quinault Indian Nation tribal officials I discovered why and how cultural and ritual practices have been modified over the years because of environmental changes that have affected indigenous lifestyles. Ronald Olson, a University of Washington researcher, wrote that even by the late 1920s when he conducted his research among the Quinault, that the fish culture was disintegrating. Other native communities have also felt the effects of cultural erosion because of changes in the environment.

The federal goal today is to protect sacred beaches and the whole natural environment, though planetary climate change is not within governmental power to control. The Quinault still claim sovereignty for their people in the management of the local resources, though they have to work with a variety of state and federal agencies, like the National Oceanic and Atmospheric Administration (NOAA), the Environmental Protection Agency (EPA), the Bureau of Indian Affairs (BIA), all with varying federal statues and regulations, and contending with judicial challenges like fishing rights.

The Quinault people have lived along the western coast of Washington State since before recorded history, and their lives have revolved around the sea from which they take their livelihood. They are a fishing people, a former whaling community. They no longer hunt whales because of new international maritime laws. Salmon, especially the blue back or sockeye salmon, is the preferred food source. However, other fish caught are trout, smelt, halibut, cod, razor clams, oysters and crabs.

Since whaling is now temporarily prohibited, the old Quinault whaling canoes still go out to sea for other fish and sometimes seals. The ocean canoes can go 12-30 miles offshore. In the old days, a chief whaler had to undergo special training that lasted one month where he was forbidden sexual intercourse and had to invoke guardian spirits to protect him and his crew. He rubbed human bones on his body and swam in the river to build up his strength. A strong whale might run after being harpooned for 10-15 miles before tiring. Occasionally the

Quinault hunt seals. But their primary food and profit source is salmon, especially the blue back salmon, and steelhead trout.

The most conspicuous change that has disrupted Quinault fishing practices has been the complete disappearance of the Anderson Glacier. The former runoff of the water from the Anderson Glacier once cooled the waters of the Quinault River. Now, the absence of the cooling glacier waters means that warmer water flows down the Quinault River. Since salmon do not normally swim in warm waters, this environmental change diminishes the legendary and life-sustaining food supply. The marine life that once maintained the community is slowly eroding, altering the food source and to some degree the cultural rituals surrounding religious ceremonies.

Traditionally, weirs were constructed at the entrance to rivers. Weirs are a series of slender poles braced to hold the river's current and erected with wood poles. A platform or two for the fishermen were erected for the men to lie across so they could look into the water. Special dip nets made of yew were lashed on two sides. Elk sinews were tied onto the upstream side of the basket-like frame. A salmon swimming upstream would hit the strings signaling to the men atop that a salmon was present.

When fishing on the ocean was more common about 25 years ago, there were as many as 28 large canoes up to 80 feet in length, plying the waves. In the past hundred years, when as many as 200,000 salmon were caught, the annual catch now is closer to 25,000.

When not fishing, men and families go into the mountain forest to hunt elk and deer. The forest is covered with dense undergrowth and impenetrable forest. It consists of Douglas fir, spruce, hemlock, and red cedar. The Quinault camp in erected shelters while the men hunt and the women pick berries and dry the meat. This process usually occurred continuously during the month of April through June when the salmon were running the strongest.

The Quinault have no concepts relating to the supernatural. They believe there are human and animal spirits in the world that have power, and that there are guardian spirits. There is no idea of worship because there are no deities. However, individuals, and sometimes families, go on spirit quests usually upriver to find a spirit power. These quests may be one day or a week or more and are equivalent to visions sought in a Navajo Hogan or a Sioux sweat lodge.

Lake Quinault used to be glacier-fed, and was glacier-carved, but the glacier, now completely gone, no longer feeds water into it. It is about eight miles long and three miles wide, and has a depth of over 1,000 feet. Moss-dripping forests and conifer trees surround the lake. A few of the oldest and largest trees in the western hemisphere exist there. The rain forest is one of the wettest places on earth with an average rainfall of between 120-180 inches, or about 10-15 feet annually. Ordinary slugs can attain a length of eight inches long.

Likewise, University of Montana researchers concluded that pollutants from five large-scale, open-pit coal mines, have leeched into the Elk River drainage in southeastern British Columbia threatening fish habitats in Canada and downstream in Washington State through the Columbia River drainage. The main pollutant is selenium, a metal-like element dangerous to fish that affects reproductive organs. This makes eating them also dangerous. Neither fish nor toxins recognize international boundaries.

Despite how humans endanger the planet and its food supply through willful indecision, indifference and neglect, rising seas are a globally recognized climate change phenomenon. As glaciers melt, warmer seas will expand. Latest estimates project that the sea level will rise 3.3 feet by 2080, flooding lower Louisiana, including New Orleans, parts of Florida and eastern North Carolina, and lower Manhattan. Indigenous peoples, especially fishing communities near the coasts, are in peril. But soon, all coastal peoples will succumb to rising seas, disrupting more than fish supplies and indigenous fishing communities.

Addressing adaptation to address climate change is going to involve multiple options and will have to include all major developed and most developing countries. America can't agree on common policies within the U.S. You can imagine how difficult it will be to conclude international agreements about a common planetary problem, especially when corporate lobbies like the Chamber of Commerce, the petroleum industries, and the National Association of Manufacturers seek to curtail efforts. The solution will also involve policies and levels of cooperation on an unprecedented scale toward a singular objective of reducing carbon emissions. New technologies will have to be employed for energy use. Lifestyle changes will be essential, and increased financing imperative. Countries in Africa, South Asia and the Middle East, and China and India, use twice as much energy as those in the developed world. On the other hand, 1.3 billion have no access to electricity. Another 2.5 billion use biomass fuels.

Energy will have to be affordable, accessible and protective of the environment. Solar and wind are the cleanest and most popular choices. Additional uses can come from geothermal, ocean thermal gradients, ocean biomass, and mineral uses like deuterium and uranium.

Not enough political will or economic investments in the best energy uses that reduce carbon emissions will occur until major natural catastrophes occur in unison, thousands die, and major coastal cities become threatened with persistent flooding. The greatest flooding threats to major U.S. cities are Miami, New Orleans, New York and Los Angeles if surges exceed three feet.

The state of Florida is truly vulnerable, as are the Florida Keys (built on the exposed debris of dying coral reefs) and Miami. Rising seas, and potentially lethal hurricanes and storms, will swamp valuable shore property unless major adaptations can be effected soon. Developers seem oblivious to these imminent dangers and continue to build in locations that are susceptible to flooding. At this writing, more than half of Florida's sandy beaches are eroding. Climate change in Florida and the predictable rising waters will have a huge impact on agriculture, tourism, and certainly the booming state of building construction.[31]

Some cities, at this writing, like Beijing and Delhi are already unbearable to breath. The air, loaded with carbon particles from China, the U.S. Brazil, Russia or wherever, does not know national boundaries and circulates freely around the earth. One country alone cannot protect the health of its citizens without a major international resolve. Prudent political choices and solid financial investments will determine whether we as a species achieve the objective of getting sufficient energy to everyone and simultaneously keep the planet safe from irreparable loss and planetary carnage.

Risky Business: The Economic Risks of Climate Change to the United States is a report by the Risky Business Project that has analyzed the economic impact of climate change on business.[32] The Risky Business Project has a leadership composed of former business and government leaders who have concluded that the business community needs to understand how climate changes will affect their bottom lines. A few of their conclusions:

- By 2050 between $66 and 106 billion worth of existing coastal property will be below sea level.

- By 2050 average temperatures will rise 2-3 times the average temperatures in the past 30 years.
- In the next few decades some states will experience a 50-70% loss of annual crop yields unless there is significant adaption.

This and other reports use a standard risk-assessment approach to determine the range of potential consequences for each region of the U.S., as well as for selected sectors of the economy. The Risky Business research focuses on the most economically significant risks, particularly changes in agricultural production and energy demand, and the impact of higher temperatures on labor productivity and public health.

The project's findings show that if America continues willy-nilly on its current path, many regions of the U.S. will face the prospect of serious economic effects. However, if we choose a different path, as the report suggests, to mitigate future impacts by reducing carbon emission, we can significantly reduce our exposure to the worst economic risks from climate change, and also demonstrate global leadership on climate, the earth's most important issue. Earth really is the Big Picture, and everything else pales in comparison.

There is increasing optimism that former mayors, Treasury Secretaries, and billionaire hedge fund managers, some of the main leaders of the Risky Business Project, might have an influence on indecisive congressional leaders.

Finally, minor business changes reveal how energy saving procedures are seeping into commercial consciousness, even though the motivation is self-interest, a very primate activity. If you have stayed in hotels or motels recently you have likely found a notice such as the following: "Dear Guest, Every day millions of gallons of water are used to wash towels that have been used only once. You make the choice—a towel on the rack means I will use it again. A towel on the floor means please replace. Thanks for helping us conserve the earth's vital resources."

It is such seemingly minor steps, if everyone would follow them that can conserve water and energy as both these resources dwindle. Future generations will have to do much more to preserve human life as we understand it.

> *This is why the earth herself, burdened and laid waste, is among the most abandoned and maltreated of our poor; she "groans in travail" (Rom 8:22). We have forgotten that we ourselves are dust of the earth (Gen 2:7); our very bodies are made up of her elements, we breathe her air and we receive life and refreshment from her waters.*[33]

The papal encyclical *Laudato Si'* of Pope Francis, released on May 24, 2015, besides describing all the scientific evidence for climate change, notes that it is the myopic mindset of people and national leaders that has failed to address the effects of climate change adequately. The encyclical is both a description of the present state of the earth and a warning about future progress. The developed world has postponed, diminished and neglected the health of the planet, and used short-term economic gain as its principal goal and not conservation of the environment. The consumer culture is the chief priority and not humanity's long-term need to survive with adequate resources. Capitalism has indeed lifted millions out of poverty, but the price has been a diminution of natural resources needed in the future. Mere technology will not be able to provide a slowdown in the more dramatic effects of climate change.

Lastly, the Greeks called her Aphrodite, the Romans named her Venus. She acquired the name given our closest solar relative, the brightest object in the night sky. The reason there is no life on Venus is because the temperature on the surface is roughly 860 degrees Fahrenheit. It is so hot there, not because Venus is closer to the sun but because the atmosphere is full of carbon dioxide that keeps the sun's heat trapped. Carbon dioxide is exactly what earthlings are pouring into our atmosphere at an unsustainable rate. You can draw your own conclusions.

The Columbia river at the California/Arizona border near needles, California.

A return to older energy options (Skagen Museum, Denmark).

Earth, the only livable planet.

EPILOGUE

I had a difficult time in my youth initially accepting my animal nature. In my religious education and training, and in various stages of immaturity, I had accepted that my humanity was a higher order, separately and divinely created in God's image to rule over the animal kingdom as prescribed in *Genesis*. My Jesuit high school had no biology or physiology course offering, just chemistry and physics, so the Jesuits who taught me were able to avoid discussing embarrassing physiological facts, so details of human anatomy remained a mystery until adulthood and my own investigations.

It was only in a collegiate anthropology class that the priest instructor suggested that the possibility might exist that God had hypothetically allowed evolution to occur. We used no textbook that might raise embarrassing questions, such as how similar we were to animals. Charles Darwin's *Origin of the Species* was not mentioned as an alternative model for the Bible's interpretation, nor was the theory of natural selection alluded to. I did not read Darwin until I was a high school teacher in 1964, and then mighty scales fell from my eyes. My Catholic education had sheltered me from the greatest scientific contribution of the ages. I felt intellectually betrayed, and by a religious order that prided itself on intellectual education.

Our human primate rule through biblical injunction, with its overpopulation and our destruction of limited natural resources—including the animal species humans reputedly have supervisory husbandry over—have illustrated how careless we have all been with our planetary and divinely-ordained stewardship. Either the divine injunction was misplaced and overly ambitious, or we have been unseemly and unsuccessful shepherds of our inheritance.

Human population has certainly learned to propagate itself, (the injunction in *Genesis* is to "increase and multiply"), nearly quadrupling in my lifetime, a disproportionate ratio that has never occurred in such a short period in human history, either with other primates or any other known animal species. Food production, thankfully, has kept pace in the developed world, but it is now orchestrated by corporate interests, and is limited and often unavailable in regions where it is most needed, contributing to malnutrition among children with all the attendant diseases accompanying a lack of balanced diets.

It is only in later life after further knowledge and study that I came to the conclusion that we were biblically misinformed, and that a religious canon—that was in reality nothing more than a series of literary documents, many of them purloined from previous literature—could be a substitute for scientific knowledge and a presumptuous prescription for the development of human life.

Unless our foundation for all knowledge is based solely on religious imperatives, we must uncomfortably accept—because there are no other validated options—that we have animal origins and a primate ancestry. We have evolved through incomprehensible epochs to emerge as human primates who have a difficult time not believing that we have a special place in the organic universe, when in fact we don't. Human primates, and indeed all primates, could become extinct as the dinosaurs once did because of an astral disaster, and the earth would simply uncaringly evolve in other ways without our presence, as it has on other planets in our solar system absent biological life. That is indeed a sobering prospect to which there is neither religious solace nor scientific predictability.

Our sense of individuality and consciousness of our personal existence leads us to believe we are immortal, an idea bolstered by the persuasions of religious doctrines, when the evidence all around us indicates the opposite. It is this traditional misperception of ourselves that deludes us into this sense of superiority.

The physical and biological sciences give us eye-opening knowledge about our evolving primate origins, the earthly milieu where we live, the ever-expanding universe with its billions of stars, quasars, and galaxies, and the archaeological findings that have made us revise human history, once based almost exclusively on the scientific inadequacies and mythologies of biblical knowledge.

The physical and biological sciences have provided the evidence, that I noted in the first couple of chapters, for the extremely old history of organic life, and the slow but ineluctable evolution over millions of years of minute organisms into large mammals and their multiple manifestations, including the many branches of the primate tree, one of which resulted in Homo Sapiens. The arrival of the genus Homo over the deep time of a few million years has been a series of experiments, of intermarriages of a variety of related hominid species, such as that between the Cro-Magnon and Neanderthal peoples about 35,000 years ago, and has not been a linear progression of one hominid type with so-called missing links.

My account of Ida, the first known primate discovered thus far at 47 million years old, and an examination of orangutans, one of our closest primates, are examples of this evolution through natural selection. The discovery of Otzi the 5,300 year-old man found in the Alps, and the 9,000 year-old Kennewick Man found in Washington State, reveals how similar we are to more recent organic primate mutations.

Based on this incontrovertible and foundational support base of humans as evolving primates, I chose to examine our human behaviors and activities through the multiple lenses of the social sciences, the primary academic methodologies for uncovering our collective identities, our behaviors, how we interact with each other in pairs and in social, economic, and political unison. In the same way that the physical sciences uphold the theory of natural selection, I believe that an examination of the human condition through its acknowledged social and scholarly fields also upholds that same theory.

I thought it necessary to examine the gender difference in greater detail because the evolution of sex as a procreation imperative is so essential to understanding most life forms. Its nearly identical organic mechanisms and use among primate populations is indicative of a common origin in mammals, and farther back in time in microorganism and multicellular construction. But more significantly, its compulsions and impulses in human life dominate so much of human activity. Sexual maturity comes earlier in females than males, but once established has a strong and pronounced influence throughout life both in individuals socially and in group behavior.

Adult maturity over time turns into senility and sliding memory loss and a childlike state of innocence prior to life's inescapable end. We mourn the passing of loved ones but should save a tear or two to acknowledge that our turn is also inevitable.

I develop the religious narrative in Chapter Four, exploring a brief background of religious developments in civilization over the last few thousand years. Revisiting some of these ancient beliefs shows how more ancient mythical writings got incorporated into later religious developments, and how Christianity as one of the world's more populous religions became an amalgam of earlier beliefs from Babylon, Persia, Egypt and Greece.

I do not believe that we inherit a specific religious belief and that what we come to believe religiously, if anything, is derived from our culture and early development formation based on the environment of our family, group and national preferences.

Closely related to the belief in transcendental realities is the belief in mental abstractions that the advent of philosophy promoted. The western intellectual heritage is that philosophy, through what the ancient Greeks left us, has empowered and enriched our culture. The concepts of mind and soul are the most influential ideas, and they strengthen and solidify religious beliefs and whole canons of intellectual persuasions. The mental benefits emanating from the genius of Plato and Aristotle, and later from some of my personal choices of Thomas Hobbes and John Stuart Mill, have become the stuff of western civilization and liberal education. I concluded the chapter on the philosophical imperative with brief descriptions of Asian and Enlightenment philosophies and the impact of Gnosticism on Christian mysticism.

We feel closest to ourselves, our bodies, our psyches, but understand them only minimally. So a study of psychology, the science of who we are, is imperative. My take is that everything we understand by our psychology can be explained by the physiology in the brain. Studies in comparative physiology—with descriptions of the work of Pavlov, Kohler and Lorenz, each of whom experimented with animal behavior, to show our behavioral proximity to our mammalian and primate cousins—confirm that primates and other animals also share similar physiognomies, impulses and emotions.

I began in the first section on psychology with a discussion of disorders because they can be visualized with new brain scanning technologies. Nearly 10% of American adults have a mood disorder, and about 6% suffer from a serious mental disorder, and half of that number have multiple and compounding mental disorders, including anxiety disorders. According to the National Alliance on Mental Illness, approximately 62 million individuals experience mental illness in a given year, and 14 million, or one in every 17, live with a serious mental illness.

My hypothesis is that these psychological disorders are rooted in the physical brain and in cellular omissions or misalignments. So, accordingly, I described the brain and neurosciences in some detail to illustrate the evolutionary path taken to arrive at this developmental stage in human progress.

The economic imperative I claim is the equivalent of primate food hoarding on a more complex scale. The abstract component is money, a substitute for actual commodities. But the human compulsion to acquire and hoard money is the same instinct animals have for getting and storing food and other necessities needed for shelter, sexual propagation and survival. I attempted to demonstrate that capitalism and the accumulation of wealth in fewer and fewer hands and pockets has generated severe income inequalities and stratified society into poorer economic conditions. The economic imperative to hoard and store is much stronger than the altruistic trait towards compassion and sharing with the group.

I selected government under the rubric of political science to illustrate how we behave in a large political context of the state. Even though through the electoral process we seem to come together to form a majority consensus on whom to represent us, we are in fact distant from most decisions that determine the life of the nation. Government too, like the complexities of the economy, is an abstract phenomenon in our lives, and we seem to relate to it directly as a herd follower would to a troop leader.

I have a peculiar position on our present political dysfunction, believing that we have always been two nations, divided more or less into North and South, with the South as the former plantation economy built on slave labor, and its subsequent century of discrimination against African Americans and to a certain degree women. Our history of slavery, White supremacy, and the residuals of anti-government sentiments rule half of our legislatures and divide us politically and socially. The agitation of the public into these dual political camps causes anxiety, passion and fervor, and contributes to tedious social discourse, because proponents on both sides chant repetitive and familiar slogans and phrases that only represent glazed-over and imperfectly understood political ideologies, like our understanding of socialism.

In Chapter Nine I explored through our political understanding the role of the Alpha male as personified by Presidents Franklin Roosevelt and Ronald Reagan. Unlike more primitive wild primates, the selection of a human Alpha male is not the winner of a physical contest but a selection process by electoral choice. I do not believe this diminishes the role of the Alpha male because the president—and other elected legislative representatives—act exactly as if they could rule the population that selected them in the same manner a troop leader might.

Finally, the unified warnings from scientists about the dangers of global warming caused by increased levels of carbon dioxide pumped into the atmosphere, can no longer be ignored, though they are in the U.S. by legislators beholden to constituencies and corporations opposed to changes. Reduction of carbon gases and methane is essential to partially forestall the negative temperature effects effecting climate change. Petroleum and energy companies like coal mining believe they have money to lose in investing in technologies that will prevent carbon emissions. Meanwhile, oil spills occur repetitively on land and sea, and coal mines explode and collapse trapping miners. The substantial campaign contributions and political clout of energy producers in several countries have curtailed political will among nations to challenge climate change with decisive green projects.

The long-term effects of climate change are unknown and not even predictable because the earth, at least as long as any records have been available, has never experienced this level of carbon and toxins in the atmosphere. And we are already experiencing its deleterious consequences in troublesome ways and in dramatic weather aberrations.

I raise this particular issue on climate change as one associated with human primate activity and in the social sciences because in the decades ahead international inaction could lead to primate extinction, or near extinction, and not just to humans. Some non-human primate extinctions, certainly among gorillas and orangutans, may occur soon because of habitat destruction.

As citizens in a national governmental structure we may appear to have miniscule ability to change anything. This lack of influence nationally is even more true in an international setting in which global cooperation is imperative.

What is new with this issue is that because of the exponential increase of humans and the increased need for food and energy, we have begun to destroy the one environment from which there is no escape. Is it possible for the collection of humans to become aware of the danger and through collective action change policies and behaviors to return the planet to some equilibrium that would be safe for future generations? I do not believe that the political courage either in the U.S. or elsewhere exists to stem the inevitable atmospheric imbalance that will literally rain more vicious and destructive weather patterns everywhere.

The earth has indeed undergone momentous weather changes in its four billion year-old history, but not while humans were on board. Human evolution has adapted to the earth's conditions only in the past few million years, more profoundly in the past few thousand years. Weather changes proceeding now could make human life unbearable, and certainly unlivable if the changes are more pronounced. But absent knowledge of the perils to the planet, this earth, because of rising numbers that Malthus first predicted, will not be kind to succeeding offspring.

And all this is in spite of the possible destruction of lives because of wars between and within nations. Already in 2015 we have been experiencing the forced migrations of over 60 million people, the greatest human movement of people since World War II, and in size equaling the approximate number killed in World War II. The breakdown of nation states into tribal, ethnic and religious identities, and into civil wars, has partially caused this mass exodus from urban and rural areas.

Innovative individuals like Stephen Hawking, Bill Gates and Elon Musk believe that the greatest danger civilization faces is from the rise of the robots and accumulating computer power that may lead to malicious intent toward humans. I think this falls in the realm of science fiction and not a prediction for future anxiety. Robotic activity and computing power will never equal the threat from climate change over the following decades.

The physical sciences illuminate where we came from, and the social sciences show us how we live, but presumably with more sense and reason than other primates. But nothing prepares us for an uncertain future where frailty, negligence, ignorance, incompetence, pollution and aggressions have shortened lives and stunted the planet's sustaining powers for untold eons. We have our individual merits and virtues. But the record of the totality of human activity is less promising, more miserable for the world's poor, more unstable in many Islamic countries, and decidedly less virtuous. We cannot roll back evolution, and its positive forward progress has defined challenges that will divide developed nations from the less developed countries, regardless of how powerful is the potential of individual abilities.

This earth, this throbbing environment that doesn't need the presence of primates to exist, will rule us all in its threatening responses to the toxicity and pollution of its enveloping atmosphere and degradation of its enveloping biosphere. Prepare for more volatile weather wherever you live because the wounded earth is responding to intemperate and careless management with its own kind of retribution. Nature isn't going to wait around while we dither trying to decide how to resolve the inflicted damage. Humans don't seem to have evolved enough as a varietal sub-species intellectually, socially or decisively to stop the destructive path. And there won't be eons of time to allow the slowness of evolution to create new super-intelligent beings who might know what to do.

This synopsis, and the book's brief exploration of the main social sciences, demonstrate how our instincts and emotions translate into ways we navigate in modern life, and constitute only one method for understanding our primate origins.

DNA and genetics rule the individual as they do all primate, mammals and organic life forms. Research in each of the social sciences—our behavioral, social and cultural responses to the environment we find ourselves in—will widen our knowledge of patterns and trends in human existence. My aim has been to propose that underlying those ongoing academic studies and investigations is the basic fact that collectively we are a rapidly populating species that owes its evolution to the whole organic development that preceded us.

Thanks for staying me with me through this intellectual argument, and indulging me in my reflective commentary and anecdotes based on my long personal and professional experiences. Sharing them, hopefully, will be a stimulus to further discussions. I hope the reading of it has been as useful to your thinking as it has been enjoyable for me in writing it.

ENDNOTES

Chapter One

[1] My first academic love as a young student was astronomy. One of my first cherished magazines I subscribed to in the 1950s was *Sky and Telescope*. My first publication was on the northern lights: "Northern Lights: Night's Rainbow." *America*, CI (8), May 23, 1959, pp. 370-72. You can download images from the Hubble telescope from several web locations. One is at: https://asunews.asu.edu/files/heic1411a.jpg. See also N. Afshordi, R. B. Mann, & R. Pourhasan, "The Black Hole at the Beginning of Time," *Scientific American*, August 2014, Vol. 311(2), pp. 37-43. For a philosophical exploration of nothing see J. Holt, *Why Does the World Exist?* (New York: Liveright Publications, 2012).

[2] K. Fedarko, "A Cathedral Under Siege," *The New York Times*, August 10, 2014, sr1ff. Major John Wesley Powell, civil war veteran buried in Arlington, led the first expedition down the Grand Canyon river gorge, and his journal descriptions are some of the first recorded about this American natural treasure.

[3] M. D. Limonick, "The Hunt for Life Beyond Earth," *National Geographic*, July 2014, pp. 26-45.

[4] E. O. Wilson, *A Window on Eternity*, *A Biologist's Walk Through Gorongosa National Park*. (New York: Simon & Schuster, 2014, p.132). For further explanation see R. Dawkins, *The Ancestor's Tale*, *A Pilgrimage to the Dawn of Life*. (Boston: Houghton Mifflin, 2004).

[5] S. J. Gould, *Wonderful Life*. (New York: W. W. Norton, 1989). Cf. also his book *The Structure of Evolutionary Theory*. (Cambridge, MA: Harvard University Press, 2002), p. 322. Gould describes the experience of discovering, at the Burgess Shale in British Columbia, 20 different arthropods from the Cambrian era. These tiny creatures lived in the sea which engulfed the earth at the time and do not exist in their present form on earth today. I benefitted greatly also from Carl Zimmer's informative book, *Evolution*, *The Triumph of an Idea*. (New York: Harper Perennial, 2006). See too N. Wade (May 5, 2015). "A New Order to Life's Origins," *The New York Times*, D4.

[6] S. C. Morris & J-B Caron, "A primitive Fish from the Cambria of North America," *Nature*, doi: 10. 1038: 13414.

[7] E. Mayr, *What Evolution Is*. (New York: Basic Books, 2001), p. 90. For more current research findings regarding the rate of evolution see: J. Zhang, & X. He, "Significant Impact of Protein Dispensability on the Instantaneous Rate of Protein Evolution." *Mol. Biol. Evol.*, April 2005. 22:1147–1155. Cf. also B. Glassey, & A. Civetta. "Positive Selection at Reproductive ADAM Genes with Potential Intercellular Binding Activity." *Mol. Biol. Evol.*, May 2004. 21:851–859.

[8] S. J. Gould, 2002. *op. cit.*, p. 37ff & 745ff. Also *ibid.*, p. 913. See P. F. Arndt, A. D. Petrov, & T. Hwa. "Distinct Changes of Genomic Biases in Nucleotide Substitution at the Time of Mammalian Radiation." *Mol. Biol. Evol.*, November, 2003. 20:1887–1896. S. J. Gould, *op.cit.*, 1989.

[9] J. A. Clack, 2005, "Getting a Leg Up on Land." *Scientific American*, 293(6), 100-107. The latest mammal of this kind was discovered in northwestern China in 2002 and the results published in *Nature* in 2006. Known as a *Guanlong Wucaii*, it is the precursor to *Tyrannosaurus Rex* and lived 160 million years ago. T. Rex lived from 85–60 million years ago or about 100 million years after *Guanlong*.

[10] Daeschler, *et.al.* A Devonian tetrapod-like fish and the evolution of the tetrapod body plan. *Nature* (April 6, 2006), 440, 757-763.

[11] N. H. Shubin, *et al.* "The pectoral fin of *Tiktaalik roseae* and the origin of the tetrapod limb." *Nature* 440, 764-771, 6 April 2006.

[12] W. Broad, "It's Sensitive. Really." *The New York Times Science*, December 13, 2005, D1.

[13] C. Dreifus, "Live Longer with Evolution? Evidence May Lie in Fruit Flies." *The New York Times*, 12/06/05, D2.

[14] R. Cano, "Amplification and sequencing of DNA from a 120-135 million-year-old weevil." *Nature*, 1993, 363, 536-7. Cf. also R. G. Crystal, "Transfer of genes to humans: early lessons and obstacles to success." *Science*, 1995, *270*, 404-410.

[15] S. Paabo, "Ancient DNA." *Scientific American*, 1993, 269 (5), 86-92. See also Kahn, P. & Gibbons, A. (1997). "DNA from an extinct human." *Science*, *277*, 176-78.

[16] S. Jones, 1993, *op. cit.*, p. 53.

[17] S. Jones, *In the Blood: God, Genes and Destiny*. London: HarperCollins, 1996.

[18] L. Thomas, 1974, *The Lives of a Cell*. New York: Bantam Books. Cf. I. Tattersall, 2001, "How We Came to be Human." *Scientific American*, *285*(6), 57-63). Robert Hooke in 1665 was looking at slivers of cork through a microscope and observed small holes between the ribbings. They reminded him of monks' quarters so he named them cells. Thus did little nothings between something become representative of the bonds within us.

[19] Another astounding insight into cellular life and its duplicative capabilities.

[20] E. O. Wilson, *A Window on Eternity, A Biologist's Walk Through Gorongosa National Park*. (New York: Simon & Schuster, 2014). For a readable account of human evolution see C. Finlayson, *The Improbable Primate, How Water Shaped Human Evolution*. (Oxford, UK: Oxford University Press, 2014). For the human primate's insignificance role in the universe, see C. Scharf, "Cosmic (In)Significance." *Scientific American*. 311(2), August 2014, 74-77.

[21] A. Gabbai, C. Floyd, "Building the New You," *Scientific American*, 46(2), May 2015, p. 41.

[22] C. Tudge & J. Young, *The Link, Uncovering Our Earliest Ancestor*. (New York: Little, Brown and Co., 2009).

[23] A. R. Wallace, *The Malay Archipelago, The Land of the Orang-Utan and the Bird of Paradise*. (London: Macmillan and Co., 1902), p.30ff. Wallace dedicated this book to Darwin.

[24] *Ibid.*, p. 47.

[25] Wood, B., "Welcome to the Family," *Scientific American*, September 2014, Vol. 311(3) pp. 43-47.

[26] "What Makes Us Special," *Scientific American*, Vol. 311(3), pp. 60-61.

[27] C. Darwin, 1859, 1979. *The Origin of the Species*. (New York: Hill & Wang). There is a library of books about evolution, and more shelves of books condemning its theory. For a partial study see books by P. J. Bowler, *Evolution, The History of an Idea*. (Berkeley, CA: The University of California Press, 1983) & C. Zimmer, *Evolution, The Triumph of an Idea*. (New York: Harper Perennial, 2006).

[28] N. Stryker, *The Thing with Feathers, The Surprising Lives of Birds And What they Reveal About Being Human*. (New York: Riverhead Books, 2014), p.148. The chicken research was conducted by the young Norwegian Thorleif Schjelderup-Ebbe and reported in his doctoral dissertation in 1921. The phrase "pecking order" begins with his research on chickens.

[29] C. Zimmer, 2006, *op. cit.*, p. 314.

Chapter Two

[1] http://www.abc.net.au/news/2014-06-11/dung-beetle-dairy-trial/5515130.

[2] E. O. Wilson, 2014, *op. cit.*, p. 75.

[3] http://news.sciencemag.org/2004/10/single-origin-eyes. Of course the earth is young compared to the 3.8 billion years of the universe. See my colleague L. M. Krauss, "A Beacon from the Big Bang," *Scientific American*, 311(4), 59-67.

[4] Even more complicated is the continual evolution of bacterium that became resistant to antibiotics, a looming global health problem. Cf. D. J. Banks, *et al.*, "Structure and Distribution of an Unusual Chimeric Genetic Element Encoding Macrolide Resistance in Phylogenetically Diverse Clones of Group A *Streptococcus." The Journal of Infectious Diseases* 2003; 188:1899-1909.

[5] According to research in *Science* magazine in 2006, researchers reported: "Our analyses suggest that on average the Neanderthal genomic sequence we obtained and the reference human genome sequence share a most recent common ancestor ~706,000 years ago, and that the human and Neanderthal ancestral populations split ~370,000 years ago, before the emergence of anatomically modern humans. Our finding that the Neanderthal and human genomes are at least 99.5% identical led us to develop and successfully implement a targeted method for recovering specific ancient DNA sequences." J. P. Noonan *et al.*, November 17, 2006, *Science*, 1113-1118. K. Wong, "The Morning of the Modern Mind." *Scientific American*, 2005, 292(6), 86-95. See also her study of Neanderthals, "Neanderthal Minds," *Scientific American*, February 2015, Vol. 312, No. 2, pp. 36-43. E. Trinkaus, *et al.* "An Early Modern Human from the Petera cu Oase, Romania." *Proceedings of the National Academy of Sciences*, 2003, 100: 11231-11236. On Neanderthal sites in Spain, cf. M. Price, "A Tale of Two Species." *ASU Research Magazine*, 46–51.

[6] J. Diamond, *The Third Chimpanzee*. (New York: HarperCollins, 1992). As Jane Goodall had noted in her early research, chimpanzees have a lethal tendency to aggression that is not influenced by humans. New research conducted over the last several years confirms this finding. Retrieved at: https:// asunews.asu.edu/ 20140918-chimpanzee-lethal-aggression.

[7] http://genome.cshlp.org/content/15/12/1746.long D. E. Lieberman, September 20, 2007. "Palaeoanthropology: Homing in on early Homo." *Nature*, 449, 291–292. One of the first studies was by W. Kohler, *The Mentality of Apes*. (New York: Harcourt, Brace, 1925). Biology Professor Robert Sapolsky spent more than 20 years living among baboons. His reflections on these experiences are in *A Primate's Memoir, A Neuroscientist's Unconventional Life Among the Baboons*. (New York: Touchstone Books, 2001). See also: http:// www.nature.com/nature/journal/v486/n7404/full/nature11128.html. R. Leakey, & R. Lewin, *Origins Reconsidered, In Search of What Makes Us Human*. (New York: Doubleday, 1992) & E. Mayr, 2001, *op. cit.*, p. 235-36, & C. Dennis, "Chimp Genome: Branching Out." *Nature*, 2005, 437, 7055, 17ff.

[8] C. Siebert, "Voice from the Deep," *Smithsonian*, June 2014, pp. 64-73.

[9] http://www.hhmi.org/research/investigators/lahn.htm.

[10] The 40% of her bones excavated came from a single bipedal female. Her brain size was about one-fourth that of a modern human. The deposits in which Lucy was found, using a 40Ar/39Ar (Argon-Argon) dating technique, indicates her age with reasonable accuracy. D. Johanson & M. Edey, *Lucy, The Beginnings of Humankind*. (New York: Simon & Schuster, 1981). Cf. also D. Johanson & L. Johanson, *Ancestors, In Search of Human Origins*. (New York: Villard Books, 1994).

[11] C. Darwin, *The Descent of Man and Selection in Relation to Sex*. (2nd Ed.). (London: John Murray, 1888), p. 2. In attempting to reconcile the God of Israel with the God of the Christians, Justin Martyr in the 2nd century CE used the term "logos spermatikos," "reasoned emissions" of God, thereby combining a physical with a transcendental reality.

[12] R. Gore, "The Rise of Mammals, Adapting, Evolving Surviving." *The National Geographic*, 203(4), 2003, pp. 2-37. For a slightly different view of primate evolution see C. Finlayson, *The Improbable Primate, How Water Shaped Human Evolution*. (Oxford, UK: Oxford University Press, 2014).

[13] R. Conniff, "Rethinking Primate Aggression." *Smithsonian*, 14(5), 2003, pp. 60-67, & R. Fouts, *Next of Kin, What Chimpanzees Have Taught Me About Who We Are*. (New York: William Morrow & Co., 1997).

[14] J. N. Wilford, "Artifacts in Africa Suggest an Earlier Modern Human," *The New York Times*, Dec. 2, 2001), p. A1.

[15] R. Gore, "The First Pioneer, A New Find Shakes the Human Tree," *National Geographic*, special edition, August 2002.

[16] S. J. Gould, October 1994. "The Evolution of Life on Earth." *Scientific American*, pp. 85-91. See also C. Zimmer, "What is a Species?" *Scientific American, 298*(6), 2008, 72–79.

[17] D. H. Erwin, July 1996. "The Mother of Mass Extinctions." *Scientific American*, pp. 72-78. Cf. P. Godefort, et al., "A Jurassic Ornithischian Dinosaur from Siberia with both Feathers and Scales." *Science* 25 July 2014: Vol. 345 no. 6195, pp. 451-455. E. Holland, "When Dinosaurs Were Doomed," *Smithsonian*, January 2015, Vol. 45(9), pp. 63-66. "Mister Big," *National Geographic*, 226(4), pp. 101-126 for text and photos of Spinosaurus.

[18] T. White, et al. "Pleistocene Homo Sapiens from Middle Awash, Ethiopia." *Nature 423*, June 12, 2003, pp. 742–747. N. Wade, "Geneticists Track More of Earliest Humans' First Itineraries." *The New York Times*, November 12, 2002, p. D1; & C. Zimmer, "How old is it?" *National Geographic*, 200(3), 2001, pp. 78-101, & Adam's Family. *The New York Times Book Review*, March 2, 2003, p. 27.

[19] P. Davies, "E.T. and God." *The Atlantic Monthly*, 292(2), 2003, pp.112-118.

[20] P. Savolainen, et al. "Genetic evidence for an East Asian origin of domestic dogs." *Science* 2002:298, 1610-1613. See also M. Shermer, "The Fossil Fallacy." *Scientific American*, 2005, 292(3) 32, & K. Lange, "Wolf to Woof, The Evolution of Dogs." *National Geographic*, 201(1), 2002, pp. 2-31. "The mutation that gives some Kermode bears a white coat is on the same gene responsible for the coloration of golden Labrador retrievers." A. Schoumatoff, *Smithsonian*, September 2015, 46(5), p. 51.

[21] C. Darwin, 1888, *op. cit.*, p. 61.

[22] E. O. Wilson, *On Human Nature*. (Cambridge, MA: Harvard University Press, 1978), & D. S. Wilson, *Darwin's Cathedral, Evolution, Religion and the Nature of Society*. (Chicago: University of Chicago Press, 2002), & two works by Ian Tattersal, "How We Came to be Human." *Scientific American*, 285(6), 2001, pp. 57-63 & *Becoming Human, Evolution and Human Uniqueness*. (New York: Harcourt Brace, 1988).

[23] K. Spindler, *The Man in the Ice*. (New York: Harmony Books, 1994).

[24] D. Preston, (September 2014). "The 9,000-year-old Man Speaks." *Smithsonian*, Vol. 46(5), pp. 52-63.

[25] *Ibid.*, p. 60.

[26] G. Hodges, "First Americans," *National Geographic*, January 2015, Vol. 227, No. 1, pp. 125-137.

[27] *Ecclesiastes* 3, 19-20.

[28] Stix, G. (September 2014). "The 'It' Factor." *Scientific American*, 311(3), pp. 72-79.

[29] *Ibid.*, p. 78.

[30] "Comparative analysis of the transcriptome across distant species" appeared in *Nature* August 28, 2014. The full list of authors can be found at: http:// www.nature.com/ nature/ journal/ v512/n7515/full/nature13424.html31. C. Walker, "First Artists," *National Geographic*, January 2015, Vol. 227, No. 1, pp. 57. I explored the cave paintings in Lascaux, Pech Merle, and the Cave of Niaux.

[32] I. Lazaridis, et al., "Ancient Human Genomes Suggest Three Ancestral Populations for Present-day Europeans," (September 18, 2014) *Nature*, 513, 409-413.

[33] T. Hobbes, *Leviathan*. (London: J. M. Dent & Sons, 1651), & Penguin Classics 1982, p. 89.

[34] B. Nye, *Undeniable, Evolution and the Science of Creation*. (New York: St. Martin's Press, 2014), p. 149.

Chapter Three

[1] The Messel pit in Germany, a UNESCO world heritage site, is the richest source of fossils from the Eocene period, ranging from 56-34 million years ago when ancient animals began to resemble modern ones, including the earliest primates like Ida. See C. Tudge & J. Young, *The Link, Uncovering our Earliest Ancestor*. (New York: Little Brown & Co., 2009). For more about the origin of sex among the earliest eukaryotic organisms, see U. Goodenough & J. Hetiman, "Origins of Eukaryotic Sexual Reproduction," *Cold Spring Harbor Perspectives in Biology*, 2014 retrieved at http://cshperspectives.cshlp.org/content/6/3/a016154.full.

[2] For a video narration see: http://www.sumanasinc.com/webcontent/animations/content/meiosis.html.

[3] J. S. Mill, (1848, 1994). *Principles of Political and Economy and Chapters on Socialism*. (New York: Oxford University Press, 1994), p. 129.

[4] S. Junger, (June 2015). "The Bonds of Battle." *Vanity Fair*, No. 658, 106-111. The American Psychological Association's *Diagnostic Statistical Manual of Mental Disorders* (DSM) added PTSD to its list of mental disorders in 1980.

[5] Charles Darwin, *The Descent of Man*. (London: John Murray, 1888), p. 47.

[6] J. Holland, "Big Fish," *National Geographic*, July 2014, p. 111.

[7] K. Tingley, June 29, 2014, "One Child, Three Parents," *The New York Times Magazine*, pp. 26ff.

[8] T. Ferris, "The Big Idea." Washington, DC: *National Geographic*, 2014, p. 70.

[9] "What Are We High on Now?" (September 8-15, 2014). *Time*, pp. 66-67.

[10] F. Nietzsche, *Beyond Good and Evil*. (Cambridge, UK: Cambridge University Press, 2002). See also D. E. Cooper, *Nietzsche*. In N. Bunnin & E. P. Tsui-James, *The Blackwell Companion to Philosophy*. (Malden, MA: Blackwell, 1996).

[11] D. A. Puts, "Beauty and the Beast: Mechanisms of Sexual Selection in Humans," *Evolution and Human Behavior*, Vol. 31(3), May 2010, pp. 157-175.

[12] B. Edgar, (September 2014). "Powers of Two." *Scientific American*, Vol. 311(3), pp. 63-67. The practice of polygamy among Mormons officially ended in 1890, though it is still unofficially practiced among a few fundamentalist Mormon communities. On monogamy among mammals see N. Stryker, 2014, p. 251. Monogamous relationships are rare in the animal world.

[13] T. R. Malthus, *An Essay on the Principle of Population*. (Homewood, ILL: Richard D. Irwin, Inc., 1963, 1798), p. 4.

[14] P. R. Ehrlich, *The Population Bomb*. (San Francisco: Sierra Club/Ballantine Books, 1968).

[15] "How to Feed the World," *Stanford Magazine*, July/August 2014, pp. 32-33. India has the highest number of undernourished people, though Africa has the highest rates of hunger and undernourishment.

[16] J. K. Bourne, "The Next Breadbasket," *National Geographic*, June 2014, pp. 47-77.

[17] J. S. Mill, *op cit*., pp. 73-74.

[18] Current updates at R. Kunzig, "Carnivore's Dilemma," *National Geographic*, November 2014, Vol. 226(5), pp. 109-135. For a longer view of the history of cattle on the world's food survival, see D. K. Sharpes, *Sacred Bull, Holy Cow, A Cultural Study of Civilization's Most Important Animal*. (New York: Peter Lang, 2006).

[19] I. Sawhill, "Beyond Marriage." *The New York Times*, September 14, 2014, sr1.

[20] P. Roberts, *The Impulse Society. America in the Age of Instant Gratification*. (New York: Bloomsbury, 2014). "Of the nearly $1.3 trillion in equity that Americans extracted from their homes between 2003 and 2005, one-third was spent on cars, boats, vacations, plasma screen TVs, and other items of personal consumption." (p. 99).

[21] W. Mischel, Ebbesen, E. B.; Raskoff Zeiss, A. (1972). "Cognitive and Attentional Mechanisms in Delay of Gratification." *Journal of Personality and Social Psychology*. 21(2), 204-218.

[22] *Too Fat to Fight*, Report by Mission: Readiness, Washington D.C., retrievedat: http:// cdn.missionreadiness.org/ MR_Too_Fat_to_Fight-1.pdf. See also "Psychosocial Dynamics of Body Weight," American Psychological Association web site retrieved at: http://www.apa.org/science/about/psa/2014/09/body-weight.aspx.

[23] J. Diamond, *The World Until Yesterday, What Can We Learn from Traditional Societies*? (New York: Viking, 2012), pp. 231ff.

Chapter Four

[1] M. T. Cicero, *On the Nature of the Gods* (trans. H. M. Poteat). (Chicago: University of Chicago Press, 1950).

[2] Boethius, *The Consolation of Philosophy*. (trans. V. E. Watts). (Baltimore, MD: Penguin Books, 1969).

[3] *Ecclesiastes, 3:19-22.* For opposing American religious viewpoints see W. Dudley, (Ed.). (2002). *Religion in America, Opposing Viewpoints*. (San Diego: Greenhaven Press, 2002).

[4] Pew Forum on Religion and Public Life, "Public Divided on Origins of Life." (Washington, DC: The Pew Research Center for the People and the Press, 2005). The Pew also reported in 2014 that 72% of Americans believed that religion was losing interest in the public square. For archaeological evidence even from fossilized primates, see C. Tudge & J. Young, *The Link, Uncovering Our Earliest Ancestor* (New York: Little, Brown and Co., 2009) cited previously.

[5] http://www.poetryfoundation.org/poem/174165.

[6] N. Cohn, *Cosmos, Chaos, and the World to Come*. (New Haven: Yale University Press, 1993), p. 50.

[7] N. K. Sandars, *The Epic of Gilgamesh*. (New York: Penguin Books, 1960).

[8] Available in full at: http://www.sacred-texts.com/ane/enuma.htm. A copy exists in the British Museum. In Egyptian folklore, Sobek on the day of creation emerged from the "Dark Waters" in order to arrange the world. The concept of creation, of making something out of nothing, is still a difficult idea because knowledge of microscopic chemical, physical and causal changes does not answer how life originated. The theory of the Big Bang of the formation of the universe is equally mysterious because one has to question where the original big blob can from.

[9] Hesiod's *Theogony. Retrieved at:* www.sunsite.berkeley.edu/OMACL/Hesiod/theogony.html.

[10] For a readable synopsis of the earliest writings on clay, a product not only of the mythology of human creation, but the actual foundation for the beginnings of all writings, see E. Chiera, *They Wrote on Clay*. (Chicago: University of Chicago Press, 1966, 1938).

[11] D. K. Sharpes, *Lords of the Scrolls, Literary Traditions in the Bible and Gospels*. (New York: Peter Lang, 2005), p.75ff. For a more humorous and playful version see of the history of religion see H. L. Mencken, *Treatise on the Gods*. (New York: Alfred A. Knopf, 1948). Albert Einsten's *Autobiographical Notes* in 1949 noted: "Through the reading of popular scientific books I soon reached the conviction that much in the stories of the Bible could not be true." In *A Stubbornly Persistent Illusion, The Essential Scientific Works of Albert Einstein*. (Edit. S. Hawking). (Philadelphia: Running Press, 2009), p. 339.

[12] Sharpes, *Ibid.*, pp. 93-104.

[13] The centrality and similarity of creation myths is also remarkable. Among the Zunis, there were two creator beings, Tawa, the Sun God, and Spider Woman, the Earth Goddess. They both helped create all living things by singing and cradling the warm clay and making humans like gods, "like you and me." G. M. Mullett, *Spider Woman Stories, Legends of the Hopi Indians*. (Tucson, AZ: University of Arizona Press, 1979).

[14] D. K. Sharpes, "Letter to the Editor" in the January 15, 2007, *U.S. News and World Report*, p. 10. The dogma proclaimed by most, but not all of the assembled bishops at the Council of Nicaea in 325, was enlarged and re-approved by the Council of Constantinople in 381.

[15] The Council of Trent (1545-1563) defined the consecrated Eucharist as *transubstantiation*, when Jesus as God purportedly enters the bread and wine at the Mass. The same philosophical term to define a divine event was used to define the union of gods, Father and Son, though, obviously, those terms presume that one is prior in existence to the other and not consubstantial. The related Greek philosophical term with *substance* is *accidents*. The reality of the bread miraculously changes, while the *accidents* of bread and wine remain the same, *accidents* as we know today consisting of atoms, protons, neutrons, electrons, and quarks. With so many *accidents* it's difficult to know where or what the *substance* is. Nevertheless, this ancient philosophical understanding is central to Catholic belief as defined by the Council of Trent. But is not in the realm of reality, nature, biology or physics. Protestants sensibly see the Eucharist ritual or communion as symbolic.

[16] D. K. Sharpes, *op. cit.*, 2005, p. 296. Based on my research and that of the majority of scholars, this is the consensus: "All of the Gospels, as we now have them, were written years after Jesus's death, and by men who had never seen Him. They show an assimilation of myths that were old before Jesus was born . . ." H. L. Mencken, *Treatise on the Gods*, (New York: Alfred F. Knopf, 1948), p. 148.

[17] D. Hume, *Dialogues Concerning Natural Religion*. (New York: Hafner Publishing Co., 1969).

[18] T. Paine, *The Age of Reason*. (New York: Gramercy Books, 1993).

[19] W. James, *The Varieties of Religious Experience*. (New York:Touchstone Book, 1997).

[20] R. Tagore. *The Religion of Man*. (London: Unwin Paperbacks, 1988). 21. B. Russell. *Religion and Science*. (New York: Oxford University Press, 1997).

Chapter Five

[1] Plato, *Republic*. (New York: Barnes & Noble). Book 5, No. 454a.
[2] E. Nahmias, "Why We Have Free Will, *Scientific American*, January 2015, Vol. 312(1), pp. 77-79.
[3] A. de Tocqueville, *Democracy in America*. (New York: Doubleday, 1969).
[4] G. Myrdal, *An American Dilemma, The Negro Problem and Modern Democracy*. (New York: Harper & Row, 1962, 1944).
[5] "I look for the resurrection of the dead and the life of the world to come."
[6] R. Descartes, *A Discourse on Method*. London: (J. M. Dent & Sons, 1637, 1941). Literary geniuses like Marcel Proust linked sensations to memory: "What we call reality is a certain relationship between these sensations and the memories which surround us at the time." M. Proust, *Remembrances of Things Past*, (New York: Random House, 1934), p. 1008.
[7] D. K. Sharpes, *Advanced Psychology for Teachers*. (New York: McGraw Hill/Primus, 1999).
[8] http://classics.mit.edu/Aristotle/politics.1.one.html.
[9] *Ibid.*
[10] In I. Edmon, *The Works of Plato*. (New York: Modern Library, 1956).
[11] T. Hobbes, *Leviathan*. (London: J. M. Dent & Sons, 1651), & in Penguin Classics, 1982.
[12] *The King's Head* pub in Galway, Ireland, is so named by the Irish executioner of Charles I, as no Englishman could be found who wanted to be the first to behead a king, but plenty of Irish volunteered.
[13] J. Micklethwait & A. Wooldridge, "The State of the State," *Foreign Affairs*, 93(4), July/August 2014, pp. 118-32.
[14] J. J. Rousseau, *The Social Contract*. (New York: Penguin Classics, 1968).
[15] T. Hobbes, *De Cive* (1642), in *Man and Citizen*. (Garden City, NY: Doubleday & Co., 1972), p. 99.
[16] *Ibid.*, p. 118.
[17] *Ibid.*, p. 119.
[18] T. Hobbes, *De Cive*, 1642, I, XIII. *De Cive, The Latin Version*. (Ed. Warrender). Oxford: At the Clarendon Press.
[19] *Leviathan*, *op. cit.*, p. 89.
[20] In another edition of *Leviathan*: T. Hobbes, *Leviathan*. (Ed. R. Tuck). (Cambridge UK: Cambridge University Press, 1991). p. 32.
[21] We might think that reason would be the obvious first choice as a criterion for electing members to legislative offices, and not just as a philosophical starting point for a shared human ability. That has not proved to be the case. It must be acknowledged that ordinary common sense among the majority of human primates has to precede the use of reason. Sadly, both seem to be in short supply among those elected to serve the interests of the commonwealth. Politicians get elected only if they represent the ideologically majority beliefs of their constituents. Reason has nothing to do with it.
[22] The references to Mill I read for this section include: Mill's "Utilitarianism." In E. A. Burtt, *The English Philosophers from Bacon to Mill*. (New York: The Modern Library, 1939), pp. 895-948; his *Autobiography*. (Ed. J. Stillinger). (Boston: Houghton Mifflin, 1969); *Essays on Politics and Society*. (Toronto: University of Toronto Press, 1977); *Principles of Political Economy, with Some of Their Applications to Social Philosophy*. (Toronto: University of Toronto Press, 1965); *Principles of Political and Economy and Chapters on Socialism*. (New York: Oxford University Press, 1994).
[23] Italics in text, Mill, *op. cit.*, p. 132. Chap. VII, 1-4 for the whole discussion of the working class. At its height tiny England ruled over Malaya, India (that once included Bangladesh and Pakistan), South Africa, Kenya, Egypt, Palestine, Cyprus, Ireland, and of course America.
[24] J. Mascaro (trans.), *Bhagavad Gita*. (New York: Penguin Press, 1962, 2003), p. 53.
[25] D. K. Sharpes, 2005, *op. cit.*, p. 183ff.
[26] Retrieved from: http://classics.mit.edu/Confucius/analects.mb.txt.
[27] There are countless sources but one historical compilation is that of R. G. Olsen, *Science and Religion, 1450-1900, From Copernicus to Darwin*. (Baltimore: The Johns Hopkins University Press, 2004).
[28] For a review of Nietzsche's thinking see *Beyond Good and Evil*. (Cambridge: Cambridge University Press, 2002); *Thus Spake Zarathustra, A Book for All and None*. (Trans. W. Kaufmann). (New York: Viking Penguin, 1966); and *Twilight of the Idols and The Anti-Christ*. (New York: Penguin Books, 1990).

Chapter Six

[1] G. Lukianff & J. Haidt, "The Coddling of the American Mind, *The Atlantic Monthly*, 316(2), pp. 42-52.
[2] Robert D. Latzman, W. D. Hopkins, A. C. Keebaugh, L. J. Young. "Personality in Chimpanzees (Pan troglodytes): Exploring the Hierarchical Structure and Associations with the Vasopressin V1A Receptor Gene." *PLoS ONE*, 2014; 9 (4): e95741 DOI: 10.1371/journal.pone.0095741. See also S. F. Anestis, "Personality and Temperament in Nonhuman Primates," *Developments in Primatology: Progress and Prospects*, 2011, pp. 169-

192 for a study of the relationship between hormones and behavior. On monkey and American anxiety see T. M. Luhrmann, "The Anxious Americans," *The New York Times*, July 19, 2015, p. Sr. 9.

[3] T. Grandin, *The Autistic Brain, Helping Different Kinds of Minds Succeed.* (Boston, Mariner Books, 2012).

[4] Kosslyn and Koenig (1992) in *Wet Mind, the New Cognitive Neuroscience* present examples of imagery based on pictures of mental imaging taken from brain scans. Their view is that "the mind is what the brain does," and that all mental activities can be understood as a description of brain functions. (1992, p. 4.) William James (1892) was quite adamant in his view that mind was nothing apart from brain when he wrote: "Mental facts cannot be properly studied apart from the physical environment of which they take cognizance." (James, 1892, p. 3.) James discounts mind as a psychological concept, and does not treat it at all in his famous psychology text from 1890. Modern neuroscientists believe that all aspects of mind can be explained materially (Crick & Koch, 1992). Gazzaniga notes from the neuroscience perspective that "most modern neurobiologists want to believe that learning and memory are somehow carried out by changes at the synaptic level." (Gazzaniga 1992, p.52). The preponderance of evidence from the biological sciences points to physiological processes causing mental activities (Gazzaniga 1992; Crick 1994). All organisms have inborn genetically programmed material which interacts with the environment. Mark Johnson even suggests that the body is in the mind (Johnson, 1987), and this is Crick's position as well (Crick, 1994). Even Baruch Spinoza, writing in the 17th century would note that ."..the mind and the body are one and the same thing, conceived at one time under the attribute of thought and at another under that of extension." (Spinoza, 1677, in the *Ethic*)

[5] D. K. Sharpes, (August 1997). "Adolescent self-concept among international native populations." Paper presented at the American Psychological Association, Chicago; "Adolescent self-concept." *Journal of Pedagogics*, (1992) 13 (1), 52-71. More in my textbook, *Advanced psychology for teachers*. (New York: McGraw Hill, 1999); see also D. Sharpes & X. Wang, "Adolescent self-concept among Han, Korean and Mongolian Chinese." *Adolescence*, (1997), 32 (128), 913-24.

[6] Jerome Kagan capitalizes on this theme in *Galen's Prophecy, Temperament in Human Nature.* (New York: Basic Books, 1994).

[7] Lewis Thomas, *The Lives of a Cell*, 1974, p. 3.

[8] S. Paabo, S. (1993). "Ancient DNA." *Scientific American*, *269* (5), 86-92.

[9] Steve Jones, *The Language of the Genes*, 1993, p. 15.

[10] Ebstein, R. P. (1996). "Dopamine D4 receptor (D4DR) exon III polymorphism associated with the human personality trait of novelty seeking." *Nature Genetics*, *12*, 78-80. Cloninger's TPQ scale measuring novelty-seeking, harm avoidance, reward dependence and persistence also were found to have a significant association with this particular dopamine receptor gene, D4DR.

[11] K. Verweij, J. H. *et al.*, "Maintenance of Genetic Variation in Human Personality: Testing Evolutionary Models by Estimating Heritability Due to Common Causal Variants and Investigating the Effect of Distant Inbreeding," *Evolution*, October 2012, Vol. 66 (10), 3238-3251. Some early research showed a strong genetic relationship to personality and behavior, conducted in part with twins reared apart done by R. Plomin: *Development, genetics and psychology*. (Hillsdale, NJ: Lawrence Erlbaum Associates, Publishers, 1986), & R. Plomin, *Genetics and experience*. (Thousand Oaks, CA: Sage Publications, 1994).

[12] http://www.thescientist.com/?articles.view/articleNo/38539/title/Male-and-Female-Brains-Wired-Differently/.

[13] http://www.scientificamerican.com/article/study-says-brains-of-gay/.

[14] http://www.webmd.com/balance/features/how-male-female-brains-differ. On similarly sized brains see E. Luders, *et al.*, (November 11, 2009). "Why Sex Matters: Brain Size Independent Difference in Gray Matter Distributions Between Men and Women," *J. of Neuroscience*, 29(45), 14265-14270.

[15] I. P. Pavlov, *Conditioned Reflexes, an Investigation of the Physiological Activity of the Cerebral Cortex.* (Oxford: Oxford University Press, 1927).

[16] J. B. Watson, *Behaviorism.* (New York: Norton, 1925), and numerous works by Skinner, including his early work *The Behavior of Organisms, an Experimental Analysis.* (New York: Appleton-Century-Crofts, 1938), and *About Behaviorism.* (New York: Knopf, 1974).

[17] W. Kohler, *The Mentality of Apes.* (London: Kegan Paul, Trench, Trubner & Co., 1925). Later research outlined how genes got transferred to humans: Crystal, R. G. (1995). "Transfer of Genes to Humans: Early Lessons and Obstacles to Success." *Science*, *270*, 404-410.

[18] *Ibid.*, p. 265.

[19] *Ibid.*, p. 152-53.

[20] *Ibid.*, p. 160.

[21] *Ibid.*, p. 36.

[22] *Ibid.*, p. 11.

[23] K. Lorenz, *On Aggression.* (New York: Harcourt, Brace & World, 1963). As an unusual example of human aggression, one professor has written about his experiences cage fighting. J. Gottschall, *The Professor in the Cage, Why Men Fight and Why We Like to Watch.* (New York: Penguin Press, 2015).

Chapter Seven

[1] T. Piketty, *Capital in the twenty-first century*. (Cambridge, MA: Belnap Press, 2014). A classic research project of the effects of capitalism over the past 300+ years from an historical perspective by a French economist.

[2] *Sallust*. (Translated by J.C. Rolfe). (Cambridge, MA: Harvard University Press, 1960), p. 22. Sallust was a Roman historian who died in 35 CE.

[3] D. Gilson, (September/October 2014). "Survival of the Richest," *Mother Jones*, 39(5), pp. 32-35. For an analysis of how the rich got richer because of government policy favoring business, see D. L. Barlett & J. B. Steele, *The Betrayal of the American Dream*, (New York: Public Affairs, 2012). Moreover, wealthy constituents have a larger voice among representatives: "All in all, our research strongly suggest that American democracy is, indeed, unequal. Members of the House of Representatives, especially Republicans, are much more responsive to their wealthier constituents than they are to their low-wealth constituents." J. H. Rhodes & B. F. Schaffner, paper retrieved at: http://people.umass.edu/schaffne/Schaffner.Rhodes.MPSA.2013.pdf.

[4] *Survey of Consumer Finances*, Federal Reserve Board, retrieved at: http:// www.federalreserve.gov/ pubs/ bulletin/ 2014/pdf/scf14.pdf. See also C. Blow, (2014, March 15). "We can't grow the gap away." *The New York Times*, A18.

[5] J. Stiglitz, *The Price of Inequality*. New York: W. W. Norton, 2013. Cf. also J. Z. Muller, (March/April 2013). "Capitalism and Inequality." *Foreign Affairs*, 92(2), 30-51.

[6] Alexis de Tocqueville, *Democracy in America*, 1956, p. 133. See Piketty, *op. cit.*, p. 327 for comparable figures for income inequality from France. We are led by political demagogues offering only unwelcoming pabulum for relief, by evangelical fanatics who suggest older superstitions, and by financial thieves harboring distrust for our ability to pay and proposing new and higher fees to empty pockets and purses.

[7] H. Tabuchi, "30,000 Lose Health Care Coverage at Walmart," *The New York Times*, October 8, 2014, p. B9. For an analysis of the unequal wealth in the U.S. see F. Zakaria, *The post-American world: Release 2.0.* (New York: W.W. Norton, 2012), p. 10. Between 1987 and 2007, funding for U.S. corrections and prisons exceeded the expenditures for higher education in 48 states, a statistic that reveals much about funding priorities and social values. More Americans voted for contestants on a TV show than voted for U.S. President.

[8] Zakaria, *op.cit.*, p.19.

[9] United Nations Development Programme. (2013). *2013 Human Development Report.* Retrieved from http://hdr.undp.org/en/2013-report. In *The Peasantry*, Honore Balzac (1799-1850) has the Abbe Brossette, after a close study of life among his parishioners, declare to his bishop: "When you see how greatly they rely on their poverty, my lord, you can guess that these peasantry are in terror of losing their great excuse for their dissolute lives." Poverty is rarely an excuse or lifestyle of choice.

[10] Zakaria, *op. cit.*, p. 56.

[11] For a history of the financial debacle see Kevin Phillips, *Bad Money, Reckless Finance, Failed Politics, and the Global Crisis of American Capitalism*. (New York: Penguin Books, 2008). The Department of Justice fined six major banks in 2015 for $6 billion dollars for criminally rigging the currency markets. In 2015 proposed budgets from the House and Senate Republican majorities—fiscal fictions that will never become law—proposed two trillion dollar whoppers, one for spending cuts and one for revenue. The claim is that adoption of these plans will drastically reduce the government's spending, mostly drastic cuts to such programs as food stamps and Medicaid and the *Affordable Care Act*. The savagery of imaginary Republican budgets seems always to be directed to the poorest in the nation. There were no details in these proposals about where new revenue was supposed to come from. Never in recent memory has unrealistic, deceitful, and some say cynical budget proposals emanated from supposedly responsible legislators. Instead, the deficit in 2013 was less than half what it was in 2009, and the economy in 2015 was slowly reviving.

[12] *Annual Report to Congress, 2013.* Washington, DC: Federal Housing Finance Agency. Retrieved at: http://www.fhfa.gov/AboutUs/Reports/Pages/Annual-Report-to-Congress-2013.aspx.

[13] D. Walker, (2014) *Quarterly Update: Foreign Ownership of US Assets*. Retrieved from http:// www.cfr.org/ staff/ b18346. Both national and personal debts are at an all-time high. 45% of GDP, or $2.2 trillion dollars, is held by foreign investors.

[14] Zakaria, *op. cit.*, p. 46.

[15] Stanford Center on Poverty and Inequality. (2014). Retrieved from: http://web.stanford.edu/group/scspi/ Consumers have been the biggest losers in the financial loss beginning in 2008, and it is partially their own fault, for not saving enough for the long-term, for believing debt was only temporary, and without any reckoning due, and for liberal, steroid use of credit cards that sponsored the lifestyle they thought they were entitled to, and therefore for living beyond incomes. Increased borrowing, or the piling on of more personal debt, fed increased consumption, the main engine of the economy, equaling, at its height, nearly 70% of the total economy.

[16] P. O'Connor, "Poll Finds Widespread Economic Anxiety," *The Wall Street Journal*, August 6, 2014, p. A4. For an analysis of the struggles of Americans, see: *Report on the Economic Well-Being of U.S. Households in 2014.* Washington, DC: Board of Governors of the Federal Reserve System. Since 1980, wages have increased

about 8% while productivity has risen 63%, noted Professor Thomas Kochan of MIT. The national goal should be to bring these two stats more into balance, and to give to workers the majority share of the nation's wealth and reduce the inequality.

[17] R. Foroohar, "Wall Street's Values are Strangling American Business," *Time*, July 21, 2014, p.14.

[18] John Stuart Mill, *Principles of Political and Economy and Chapters on Socialism*. (New York: Oxford University Press, 1994), p. 316.

[19] *Op. cit.*, p. 269.

[20] C. Hagel, *America, Our Next Chapter*. (New York: Ecco Press, 2014). p. 12. For a future perspective on work see D. Thompson, (July/August 2015). "A World Without Work," *The Atlantic*, 316(1), 51-61.

[21] C. D. Goldin & L. F. Katz. *The Race Between Education and Technology: The Evolution of U.S. Educational Wage Differentials, 1890-2005*. (Cambridge, MA: The Belnap Press, 2010).

[22] Piketty, *op. cit.*, p. 335. And now, through conservative political moves, parents have more decisions about the schools their children attend, there is increased segregation and less school population diversity, the antithesis of what we supposedly stand for in our democracy. See H. Knight, "Racial Islands in San Francisco's Sea of Diversity," *San Francisco Chronicle*, May 17, 2015, Section R.

[23] Piketty, *op.cit.*, p. 438.

[24] J. S. Mill, *op. cit.*, p. 127.

[25] T. McMillan, "The New Face of Hunger," *Scientific American*, August 2014, 226(2), pp. 67-84. Cf. also L. Panetta, *Worthy Fights*. (New York: Penguin Books, 2014), p. 75.

[26] For a further analysis of the Peasants' Revolt, see Barbara Tuchman, *A Distant Mirror, The Calamitous 14th Century*. (New York: Alfred Knopf, 1978), pp. 372-78.

[27] For historical background on the origins of American capitalism see M. Josephson, *The Robber Barons, The Great American Capitalists, 1861-1901*. (New York: Harcourt, Brace and Co., 1934), & C. R. Morris, *The Tycoons*. (New York: Henry Holt & Co., 2005).

[28] K.Q. Seelye, "A Mother Lifts Her Son, Slowly, From Heroin's Abyss," *The New York Times*, August 11, 2014, p. A9. It's unknown whether or not there is a casual relationship between a depressed economy and a search for some substitute pleasure, like the Freudian pleasure principle of drugs, to remove the disappointing pain. I know of no evidence, but suspect there is a possible correlation. A government crackdown on prescription drug abuse has led many to use heroin, cheaper, but much more damaging.

[29] Piketty, *op. cit.*, p. 475. But there are also high and highly disparate tax revenues to support the state's role in satisfying citizen needs. A hundred years ago in 1910, tax revenues for Sweden, Britain, France and the U.S. all hovered around 10%. By 2010 there were wide disparities. Sweden and most Scandinavian countries were around 55%, France was at 50%, Britain at 40% and the laggard U.S. at 32%. [29] At the turn of the 20th century developed states had little involvement in the economic and social life of the people. Today, the rise of the so-called socialist state has occurred because states have agreed to redistribute tax revenues—in addition to essential safety services like fire and police protection—to health, education and pensions based on citizen demands. This is not a redistribution of wealth from the rich to the poor, but to all citizens.

[30] *Ibid.*, p. 510.

[31] J. E. Stiglitz, (June 29, 2014), "Inequality is Not Inevitable." *The New York Times*, sr1. Ask yourself if it can't be the responsibility of a modern society to pay for elections with public funds, to have strict gun laws, to sponsor child care so parents can raise families without excessive burdens, to have a defined percentage of all revenues devoted to education at all levels, including vocational training related to jobs, and to have guaranteed lifetime health care. Is any of this socialism, or just good government? Then ask yourself whether or not private industry does any of these public services exclusively for the commonweal?

[32] The judicially misguided U.S. Supreme Court decision, *Burwell v. Hobby Lobby Stores*, June 30, 2014, that allowed owners not to provide contraceptive health insurance coverage under the *Affordable Health Care Act* if it was against the religious beliefs of the owners. Even the U.S. Supreme Court favored the religious beliefs of businesses against the general welfare of workers. Are not these corporations, and other businesses like them, exactly the people we want to have all the individual rights of freedom other citizens have? And are not these the same corporations we want unregulated so they can make all the money they want but cheat their consumers, pollute the environment if they can, and retailers that don't spend enough to protect credit card purchasers from hackers?

[33] *193 U.S. 197* (1904). This law was used in 2000 to deny Microsoft its web browser monopoly control.

[34] Teddy Roosevelt's anger upon reading Upton Sinclair's *The Jungle* in 1906 prompted Congress to pass legislation to Congress. Still, corporate meat and chicken producers occasionally are faulted with Ecoli and salmonella contaminations because of carelessness and neglect of public safety standards.

[35] F. Fukuyama, (September/October 2014). "America in Decay, The Sources of Political Dysfunction." *Foreign Affairs*, 93(5), pp. 5-26, and see similar expressions in my article, D. K. Sharpes. (August 2014). "American Economic Decline: A Consequence of Lower Educational Standards." *US-China Education Review*, Vol. 4(8), pp. 560-66.

[36] http://www.gao.gov/fiscal_outlook/federal_fiscal_outlook/overview.

[37] R. H. Tawney, *The Acquisitive Society*, (New York: Harcourt, Brace & Co., no date but c. 1926), p. 184. In 1952 the share of corporate federal tax revenue was 33%. In 2014 that same share had dwindled to 10.5% at a time when corporate profits were at $1.8 trillion for the second quarter of 2014. Profits in 2014 were up 21% while America's total corporate tax bill had declined to 5%. The big waves from speeding corporate yachts are sinking the canoes and rowboats of everyone else.

Chapter Eight

[1] Adam Smith, *The Wealth of Nations*, Book II, 2. 94.

[2] *The New Yorker*, October 27, 2014, pp. 19-20.

[3] Panetta, *Worthy Fights*. (New York: Penguin Press, 2014), p. 422.

[4] D. Sharpes, *Our Divided House, Anti-Government Movements and the Failure of Unification*. (Vulcan: CreateSpace, a Division of Amazon, 2012). For the modern version of congressional division see T. Davis, M. Frost, R. Cohen, *The Partisan Divide, Congress in Crisis*. (Campbell, CA: Premiere, 2014).

[5] Those ignorant of the Constitution always cite the 10th Amendment as indicating that states have something similar to abdication of whatever the federal government proposes. They never seem to have read Article VI of the Constitution where it is abundantly clear that congressional legislation is sovereign over any legislation by states. Article VI declares: "*This Constitution, and the laws of the United States which shall be made in pursuance thereof . . . shall be the* ***supreme law of the land****, and the Judges in every state shall be bound thereby, anything in the Constitution* ***or laws of any State*** *to the contrary notwithstanding.*" (Article VI, 2)

[6] R. Hofstadter *Anti-Intellectualism in American Life*. (New York: Vintage Books, 1963), p. 431.

[7] See the U.S. Supreme Court decision *Texas v White*, 74 U.S. 700, 1869, that belatedly prohibited secession.

[8] T. Frank, "What's On My Ballot," (September 2014) Phoenix: State of Arizona Publicity Pamphlet, p. 21.

[9] A. DeTocqueville, *Democracy in America*, 1956, p. 137. For a more thorough exploration of slavery in the U.S. consult G. Myrdal, *An American Dilemma, The Negro Problem and Modern Democracy*. (New York: Harper & Row, 1962, 1944), a classic study by the renowned Swedish sociologist.

[10] John Quincy Adams in Kaplan, 2014, p. 484. President Kennedy immortalized the courage of Adams in *Profiles in Courage*. (New York: Harper & Row, 1964).

[11] G. Packer, *The Unwinding, An Inner History of the New America*. (New York: Farrar, Straus and Giroux, 2013).

[12] A case made by others familiar with the Constitution, such as Peter Segal, host of "Constitution USA with Peter Segal" on PBS. See "Slaves and the Constitution," *The New York Times*, September 18, 2015, p. A24.

[13] F. Fukuyama, "America in Decay, The Sources of Political Dysfunction." *Foreign Affairs*, September/October 2014), 93(5), p 16.

[14] *Mother Jones*, November/December 2014, pp. 46-47.

Chapter Nine

[1] B. Boyarsky, *Ronald Reagan, His Life and Rise to the Presidency*. (New York: Random House, 1981), p. 25.

[2] R. Hofstadter, *Anti-Intellectualism in American Life*. (New York: Vintage Books, 1963), p. 431. Cf. also R. Jenkins, *Franklin Delano Roosevelt*. (New York: Times Books, Henry Holt, 2003).

[3] A. M. Schlesinger, Jr., *The Coming of the New Deal*. (Boston: Houghton Mifflin Co., 1958), pp. 297-315.

[4] Retrieved at: http://millercenter.org/president/speeches/speech-3320.

[5] *The Economic Conditions of the South*. (Washington, DC: U.S. Government Printing Office, issued June 22, 1938).

[6] J. M. Burns, *Roosevelt: The Lion and the Fox*. (New York: A Harvest/HBJ Book, 1956), p. 85. Cf. also R. Jenkins, *Franklin Delano Roosevelt*. (New York: Times Books, Henry Holt, 2003). For a pictorial history of the Civilian Conservation Corps, one of the most successful of Roosevelt's government programs that put people to work see S. Cohen, *The Tree Army, A Pictorial History of the Civilian Conservation Corps, 1933-1942*. (Missoula. MT: Pictorial Histories Publishing Co., 1980).

[7] S. Kennedy, *Southern Exposure*. (Garden City, NY: Doubleday & Co., 1946). In 1936 one of the houses in Washington State I lived in as a child next to the railroad tracks had no phone, no electricity (we used kerosene lamps at night), and no indoor toilet but an outhouse. The town's Blacks lived in a restricted area near the river.

[8] Roosevelt's famous phrase in the 1933 inaugural address—"The only thing we have to fear is, fear itself"—was said to have come from Emerson, whom Roosevelt probably read while at Harvard. But it is also a phrase found much earlier in Seneca, counselor to the emperor Nero.

[9] I relied chiefly on two biographies of Reagan: B. Boyarsky, *Ronald Reagan, His Life and Rise to the Presidency*. (New York: Random House, 1981), & H. W. Brands, *Reagan, The Life*. (New York: Doubleday, 2015). I am also grateful for a small grant from the Emeritus College at Arizona State University that allowed me to spend time at the Reagan Presidential Library to gain access to documents on domestic policies.

[10] R. Maddow, *Drift, The Unmooring of American Military power*, (New York: Crown Publishers, 2012), pp. 46-69.

[11] Reagan's plan violated at least two congressional laws: *The Arms Export Control Act*, prohibiting secret arms transfers, and *The Foreign Assistance Act* that forbade the CIA from undertaking covert operations in a foreign country without a presidential finding. As it was, Reagan never relied on evidence or data to make his momentous decisions, and at the end of his presidency, in the early stages of Alzheimer's, sadly could not remember what he had agreed to or decided. See H. W. Brands, *op. cit.*, p. 543.

[12] As noted earlier, I located all these facts at the Ronald Reagan presidential library archives.

[13] Brands, *ibid.*, p. 706.

Chapter Ten

[1] For a scientific analysis of climate change on sea life see C. Roberts. *The Ocean of Life, The Fate of Man and the Sea*. (New York: Penguin Books, 2012). For an exploration of denials about climate change see C. C. Mann, (September 2014). "How to Talk About Climate Change So People will Listen." The Atlantic, Vol. 314(2), pp. 86-99. One wag has coined the term "bozone" for the substance surrounding stupid people that keeps new ideas from penetrating. http://www.ipcc.ch/report/graphics/index.php?t=Special% 20Reports&r= SRCCS&f= Technical%20Summary. Other reports of the Intergovernmental Panel on Climate Change (IPCC) are available on the IPCC's website. See also AAAS *Climate Science Panel Report*, March 18, 2014 at http://whatweknow.aaas.org/

[2] For those interested in further reading about its watery demise I recommend "Sins of the Aral Sea" in the *National Geographic magazine* for June 2015.

[3] Pope Francis retrieved at: http://w2.vatican.va/content/francesco/en/encyclicals/documents/papa-francesco_20150524_enciclica-laudato-si.html.

[4] E. Lipton, "Energy and Regulators on One Team." *The New York Times*, (December 7, 2014), A1, 30-31.

[5] M. Connelly, (September 23, 2014). "Global Warming Concerns Grow." *The New York Times*, D3. Climate change should not be a partisan issue, but it is. But on climate change, because they are beholden to corporate energy companies, legislators beholden to lobbyists seek to cut regulations that control carbon emissions, and retard the efforts of the Environmental Protection Agency whose legislation (advanced by Republican President Nixon) was explicitly created to protect the air and water quality of all Americans.

[6] E. O. Hulbert, "The Temperature of the Lower Atmosphere of the Earth," *Physical Review*, November 15, 1931, 38(1876).

[7] See several reports in the journal *Nature Geoscience* retrieved at: http://www.nature.com/ngeo/index.html. Greenhouse gas emissions have harmed the oceans and in turn have melted snow packs. If melting trends continue in Greenland and in places like Alaska and Glacier National Park, Glacier may have to be re-named Non-Glacier Park. See M. Wines, "Climate Change Threatens to Strip the Identity of Glacier National Park," *The New York Times*, November 23, 2014, p. 17.

[8] Retrieved at: www.ipcc.ch

[9] Ibid. For carbon emission see Asefi-Najafabady, et al. "A Multiyear, Global Gridded Fossil Fuel CO2 Emission Data Product: Evaluation and Analysis of Results." *J. of Geophysical Research*. Retrieved at: http://onlinelibrary.wiley.com/doi/10.1002/2013JD021296/abstract.

[10] Another irony is that scores of perfectly preserved human primates have been uncovered in peat bogs in Ireland, England, Denmark and northern Germany, and I have studied two of them, Tollund Man (c. 400-300 CE) and Grauballe Man (c. 290 BCE) found in the vicinity of Aarhus, Denmark. So humans have been interred in our fuel sources for centuries. Anatomical investigations of their preserved bodies have revealed much about our ancestral past.

[11] N. Angier, "Basics: Too Much of a Good Thing." *The New York Times*, September 23, 2014, D2.

[12] Retrieved at: http://www.ipcc.ch/

[13] The following primary sources come from my research of the presidential White House staff papers in the Ronald Reagan presidential library in Simi Valley, California. This quote is from the Elizabeth Dole's staff papers, Box 19 (5 of 9), August 30, 1981.

[14] Elizabeth Dole's memo to Chief of Staff James Baker, Box 19 (6 of 9), October 15, 1981. See also *Public Attitudes Toward the Clean Air Act*. Report of the Chamber of Commerce of the United States. (Princeton, December 1981). See too, H. Waxman, "What the Auto Industry Wants to Do to the Clean Air Act," *The New York Times*, March 22, 1982.

[15] *Contra Costa Times*, August 6, 2015.

[16] Danny Boggs papers, retrieved from OA11477, Box 70, Reagan presidential library.

[17] Staff Papers, Judy Black, "Clean Water Act," OA16830, Reagan presidential library.

[18] E. Dole staff papers, Box 19, Reagan presidential library. See E. Randolph, "Mine Accident Rise Linked to Reagan Policy, *The Los Angeles Times*, January 25, 1982. Elizabeth Dole was the wife of former Senator and presidential candidate Bob Dole.

[19] Robert Carleson papers at OA10554, Box 19.

[20] K. Galbriath, "Tapped Out," *The Stanford Magazine*, July/August, 2014, pp. 42-49. For an analysis of the metrics of water measurement and the difference between ground water and sources underground, see A. Lustgarten, "How the West Overcounts Its Water," *The New York Times*, July 19, 2015, p. Sr.1 & Sr. 6.

[21] C. Vaughn, "Land of Perpetual Drought," *Hi-Desert Star*, Yucca Valley, CA, February 12, 2014, p. A1. As fields dry out and grasses become tinder, wildfires and the threat of wildfires worsen.

[22] M. Nijhuis, (October 2014). "When the Snows Fail." *National Geographic*, 226(4), pp. 56-77.

[23] A. Nagourney, "Across Rural California, Barely a Drop to Drink," *The New York Times*, March 8, 2014, A1. California had to impose mandatory cuts on water usage. K. Fagan, "Facing Drought's Hard Reality, *San Francisco Chronicle*, April 19, 2015, A1.

[24] J. McWilliams, "Meat Makes the Planet Thirsty," *The New York Times*, March 8, 2014, A19.

[25] B. Roberts & E. J. Kytle, "The Dust Bowl Returns," *The New York Times*, February 10, 2014, A19.

[26] National Audubon Society Report Retrieved at: http://climate.audubon.org/article/audubon-report-glance#FAQ-2.

[27] "A Vanishing Migration," *National Geographic*, November 2014, Vol. 226(5), p. 22.

[28] A. Einstein, "The Foundation of the General Theory of Relativity," in *A Stubbornly Persistent Illusion, The Essential Scientific Works of Albert Einstein*. (Edit. S. Hawking). (Philadelphia: Running Press, 2009), p. 83.

[29] J. Gillis, "Sun and Wind Transforming Global Landscape," *The New York Times*, September 14, A1ff.

[30] D. Sharpes, "The Quinault Tribal Fishing Community." *Emeritus Voices*, 2013, Vol. 13, pp. 53-57.

[31] L. Parker, "Treading Water," *National Geographic*, February 2015, Vol. 227, No. 2, pp. 106-125.

[32] http://riskybusiness.org/reports/national-report/executive-summary.

[33] Pope Francis, *Laudato Si'* encyclical, May 24, 2015.

References

Achenbach, J. (March 2105). "The Age of Disbelief." *National Geographic*, Vol. *227*(3), pp. 30-47.

Adams, J. Q. *State of the Union Address*, December 6, 1825. *Report of the Minority of the Committee of Manufacturers, Submitted to the House of Representatives of the United States*. Boston: John Eastburn, 1833.

Adkins, L. & R. (1991). *Introduction to the Romans*. Secaucus, NJ: Chartwell Books.

Adler, A. (1964). *Social Interest: A Challenge to Mankind*. New York: Capricorn Books. (1969). *The Science of Living*. New York: Doubleday.

Adler, M. (1981). *Six Great Ideas*. New York: Collier Books.

____(1982). *The Paidaeia Proposal*. New York: Macmillan.

Afshordi, N., Mann, R. B. & Pourhasan, R. "The Black Hole at the Beginning of Time." *Scientific American*, August 2014, Vo. *311*(2), 37-43.

Alasuietari, P. & L. Bichman, J. Brannen. (2008). *The Sage Handbook of Social Research Methods*. Los Angeles: Sage Publications.

Allport, G. W. (1958). *The Nature of Prejudice*. New York: Doubleday.

Allport, G.W. (1937). *Personality: A Psychological Interpretation*. New York: Holt.

Anchor, R. (1979, 1967). *The Enlightenment Tradition*. Berkeley, CA: University of California Press. *Annual Report to Congress, 2013*. Washington, DC: Federal Housing Finance Agency. Retrieved at: http://www.fhfa.gov/AboutUs/Reports/Pages/Annual-Report-to-Congress-2013.aspx.

Anselm (1962). *St. Anselm Basic Writings*. LaSalle, Ill: Open Court. (1909). *Cur Deus Homo*. Edinburgh: John Grant.

Aquinas, T. (1920). *The Summa Theologica* (V. 1), London: Burns, Oates & Washbourne.

Arendt, H. (1978). *The Life of the Mind*. New York: Harcourt.

Aristotle (1975). *The Nicomochean Ethics*. (tr. David Ross), London: Oxford University Press.

____(1984). *The Complete Works of Aristotle, The Revised Oxford Translation*. Princeton: Princeton University Press. (1988). *The Politics*. (Ed. S. Everson). New York: Cambridge University Press.

Arndt, P. F., Petrov, A. D. & Hwa, T. (2003, November). "Distinct Changes Of Genomic Biases in Nucleotide Substitution at the Time of Mammalian Radiation." *Molecular Biology and Evolution*, *20*, 1887–1896.

Ascham, R. (1570). *The Scholemaster*. London: John Daye.

Asefi-Najafabady, S. P. Rayner, J., Gurney, K. R., McRobert, A., Song, Y., Coltin, K., Huang, J., Elvidge, C. & Baugh, K. "A Multiyear, Global Gridded Fossil Fuel CO2 Emission Data Product: Evaluation and Analysis of Results." *J. of Geophysical Research.* Retrieved at: http://onlinelibrary.wiley.com/doi/10.1002/2013JD021296/abstract.

Ashton, M. (1968). *The Fifteenth Century: The Prospect of Europe*. New York: W. W. Norton.

Augustine, (1960). *The Confessions*. Garden City, NJ: Doubleday.

—— (1950). *The City of God*. New York: Modern Library.

Aurelius, M. (1952). *Meditations*. In R. M. Hutchins (Ed.), *Great Books of the Western World*. Chicago: Encyclopaedia Britannica.

Babbie, E. R. (1979). *The Practice of Social Research*. Belmont, CA: Wadsworth.

Backhouse, J. (1981). *The Lindisfarne Gospels*. Oxford: Phaidon.

Bacon, F. (1955). *Selected Writings of Francis Bacon*. New York: The Modern Library.

____(1944). *The Advancement of Learning and New Atlantis*. London: Oxford University Press.

____(1853). *The Essays, or Counsels Civil and Moral*. London: John Parker and Son.

Bacon, R. (1683). *Speculum Alchymiae, The True Glass of Alchemy*. London: William Cooper.

____(1928). *Opus Majus*. (Trans. B. Burke). Philadelphia: University of Pennsylvania Press.

Bailyn, B. (1992). *The Ideological Origins of the American Revolution*. Cambridge: Harvard University Press.

Balz-ac. H. (no date). *The Works of Honore de Balzac*. V. 12. New York: Heart's International Library Co.

Barker, E. (1960). *Social Contract*. New York: Oxford University Press.

Barkow, J. H., Cosmides, L. & Tooby, J. (1992). *The Adapted Mind, Evolutionary Psychology and the Generation of Culture*. New York: Oxford University Press.

Barnes, J. (1995). *The Cambridge Companion to Aristotle*. Cambridge: Cambridge University Press.

Barnett, S. A. & Fuller, S. (1998). Science. *Interdisciplinary Science Reviews*, *23* (4), 331-39.

Barone, E. (September 8-15, 2014). "Who We Are." *Time*, 53-58.

Barlett, D. L. & J. B. Steele. (2012). *The Betrayal of the American Dream*. New York: Public Affairs.

Barzun, J. (1983). *A Stroll with William James*. New York: Harper & Row.

—— (2000). *From Dawn to Decadence, 1500 to the Present*. New York: Perennial.

____ (1989). *The Culture We Deserve*. Middletown, CT: Wesleyan University Press.

Beard, C. A. (1913, 2004). *An Economic Interpretation of the Constitution of the United States*. New York: Dover Publications.

Bentham, J. (1939). *An Introduction to the Principles of Morals and Legislation*. In Burtt, E. A. *The English Philosophers from Bacon to Mill*. New York: The Modern Library, 791-856.

Bergin, T. G. & Fisch, M. H. (1948). *The New Science of Giambattista Vico* (Trans from 3rd Ed., 1744). Ithaca, NY: Cornell University Press.

Bergman, C. (1996). *Orion's Legacy, A Cultural History of Man as Hunter*. New York: A Dutton Book.

Berlin, I. (1997). The concept of scientific history. In I. Berlin, *The Proper Study of Mankind*. New York: Farrar, Strauss and Giroux.

____(1977). *Vico and Herder, Two Studies in the History of Ideas*. NewYork: Vintage Books.

____(1997). *The Proper Study of Mankind, An Anthology of Essays*. New York: Farrar, Strauss and Giroux.

Bestor, A. (1956). *The Restoration of Learning, a Program for Redeeming the Promise of American Education*. New York: Alfred Knopf.

Bickford, C. B. & Bowling, K. R. (1989). *Birth of the Nation: The First Federal Congress 1789-1791*. Madison, WI: Madison House.

Binet, A. (1977, 1890). *Alterations of Personality*. Washington, DC: University Publications of America.

—— (1905, 1916). The Development of Intelligence in Children. In D. N. Robinson (Ed.), (1977). *Significant Contributions to the History of Psychology, 1750-1920*. Washington, DC: University Publications of America.

Birley, A. (1993, 1966). *Marcus Aurelius*. London: B. T. Batsford.

Blades, W. (1861). *The Life and Typography of William Caxton, England's First Printer*. London: Joseph Lilly.

Block, S. S. (1975). *Benjamin Franklin, His Wit, Wisdom and Women*. New York: Hasting House.

Bloom, A. (1987). *The Closing of the American Mind*. New York: Simon & Schuster.

Blow, C. (2014, March 15). "We can't grow the gap away." *The New York Times*, A18.

Bock, K. (1994). *Human Nature Mythology*. Urbana: University of Illinois Press.

Boethius. (1969). *The Consolation of Philosophy*. (Watts, V. E., trans). New York: Penguin Books.

Bonaventure. (1993). *The Journey of the Mind to God*. Indianapolis: Hackett Publishing.

Boorstin, D. J. (1976). *The Exploring Spirit, America and the World Experience*. London: British Broadcasting Corporation.

____(1983). *The Discoverers*. New York: Random House.

——(1992). *The Creators*. New York: Random House.

Bouchard, T. J., Lykken, D. T., McGue, M., Segal, N. L. & Tellegen, A. (1990). "Sources of human psychological differences: The Minnesota study of twins reared apart." *Science*, 223-28.

Borchard, R. (1957). *John Stuart Mill, the Man*. London: Watts.

Bowen, J. (1972). *A History of Western Education, Vol. 1, The Ancient World*. London: Metheun & Co.

Bowen, J. (1975). *A History of Western Education, Vol. 2, The Civilization of Europe*. London: Metheun & Co.

Bowers, C. (1969). *The Young Jefferson, 1743-1789*. New York: Houghton Mifflin. (1967). *Jefferson in Power*. Cambridge, MA: Houghton Mifflin.

Bowler, P. J. (1983). *Evolution, The History of an Idea*. Berkeley, CA: The University of California Press.

Boyarsky, B. (1981). *Ronald Reagan, His Life and Rise to the Presidency*. New York: Random House.

Boyd, J. P. (1955). *The Papers of Thomas Jefferson*. Princeton: Princeton University Press.

Boyd, W. (1963, 1911). *The Educational Theory of Jean Jacques Rousseau*. New York: Russell & Russell.

Boyd, W. & King, E. J. (1975, 1921). *The History of Western Education*. London: Adam and Charles Black.

Boyer, B. & McKeon, R. (1976). *Peter Abelard, Sic et Non, a Critical Edition.* (v. 1-6). Chicago: University of Chicago Press.

Brandt, I. (1965). *The Bill of Rights, Its Origins and Meaning.* Indianapolis: Bobbs Merrill.

Brands, H. W. (2015). *Reagan, The Life.* New York: Doubleday.

Brewer, J. S. (1859). *Fr. Rogeri Bacon, opera quaedam hactenus inedita.* London: Longman, Green, Longman, & Roberts.

Bridges, J. H. (1914). *The Life and Work of Roger Bacon.* London: Williams & Norgate.

Brodie, F. (1974). *Thomas Jefferson, An Intimate History.* New York: W. W. Norton.

Bronowski, J. (1956). *Science and Human Values.* New York: Julian Messner.

____(1973). *The Ascent of Man.* Boston: Little, Brown.

Bronowski, J. & Mazlish, B. (1960). *The Western Intellectual Tradition.* New York: Harper & Brothers.

Brown, M. P. (1998). *The British Library Guide to Writing and Scripts, History and Techniques.* London: The British Library.

Brown, R. (1995). *Managing the Learning of History.* London: David Fulton.

Browne, J. (1995). *Charles Darwin.* London: Jonathan Cape.

Buchan, J. (2006). *Adam Smith, and the Pursuit of Perfect Liberty.* London: Profile Books.

Bullock, A. (1985). *The Humanist Tradition in the West.* New York: W. W. Norton.

Burke, P. (1997). *Varieties of Cultural History.* Ithaca, NY: Cornell University Press.

—— (1985). *Vico.* Oxford: Oxford University Press.

Burn, L. (1990). *Greek myths.* London: The British Museum.

Burtt, E. A. (1939). *The English Philosophers from Bacon to Mill.* New York: The Modern Library.

Bury, J. B. & Cook, S. A. & Peacock, F. E. (1925). *The Cambridge Ancient History*, Vol. *3*. New York: Macmillan.

Butterfield, H. (1957). *The Origins of Modern Science, 1300-1800.* New York: The Free Press.

Cahill, T. (1995). *How the Irish Saved Civilization, the Untold Story of Ireland's Heroic Role from the Fall of Rome to the Rise of Medieval Europe.* New York: Anchor Books.

—— (2003). *Sailing the Wine-Dark Sea, Why the Greeks Matter.* New York: Nan A. Talese/Doubleday.

Calvin, W. H. (1990). *The Ascent of Mind.* New York: Bantam Books.

Campbell, J. (1988). *Joseph Campbell, the Power of Myth* (with Bill Moyers) New York: Doubleday.

—— (1987). *The Masks of God: Primitive Mythology.* New York: Penguin.

—— (ed.) (1955). *The Mysteries, Papers from the Eranos Yearbooks.* Princeton: Princeton University Press, Bollingen Series, V2.

—— (1990). *The Hero's Journey, Joseph Campbell on his Life and Work.* San Francisco: HarperSanFrancisco.

Campbell, D. T. & Stanley, J. C. (1963). Experimental and Quasi-Experimental designs for research. In N. L. Gage (Ed.) *Handbook for research on teaching.* Chicago: Rand McNally.

Cano, R. (1993). Amplification and Sequencing of DNA from a 120–135 million-year-old weevil. *Nature*, *363*, 536–7.

Cantor, N. F. (1993). *The Civilization of the Middle Ages.* New York: HarperPerennial.

—— N. F. (1991). *Inventing the Middle Ages.* Cambridge: Lutterworth Press.

Capaldi, N. (2004). *John Stuart Mill, A Biography*. Cambridge: Cambridge University Press.

Carroll, S., B. B. Prud'homme, & N. Gompel. (May 2008). "Regulating Evolution." *Scientific American, 298*(5), 61–67.

Carson, R. (1962). *The Silent Spring*. Boston: Houghton Mifflin.

Cattell, R. B. (1965). *The Scientific Basis of Personality*. Chicago: Aldin Publishing.

—— (1966). *Handbook of multivariate experimental psychology*. Chicago: Rand McNally.

—— (1971). *Abilities, Their Structure, Growth, and Action*. Boston: Houghton Mifflin.

——(1966). "The meaning and strategic use of factor analysis." (Ed.) *Handbook of Multivariate Experimental Psychology*. Chicago: Rand McNally, 174-243.

Catto, J. I. & Evans, R. (1992). *The History of the University of Oxford*. Oxford: The Clarendon Press.

Chesterton, G. K. (1933). *St. Thomas Aquinas*. London: Hodder & Stoughton.

—— (1957, 1924). *St. Francis of Assisi*. Garden City, NY: Image Books, Doubleday.

Churchill, W. S. (1956). *A History of the English-Speaking Peoples*, V. 2. London: Cassell and Co.

Cicero, (1951). *The Basic works of Cicero*. (Hadas, M. Ed.) New York: The Modern Library.

—— (1892). *Life and letters*. Edinburgh: W. P. Nimmo, Hay, & Mitchell.

Clack, J. A. (2005). "Getting a Leg Up on Land." *Scientific American, 293*(6), 100–107.

Clark, K. (1969). *Civilisation, a Personal View*. London: Penguin Books.

Clark, J. (1989). *Saxon and Norman London*. London: Museum of London.

Clegg, B. (2003). *The First Scientist, A Life of Roger Bacon*. London: Constable.

Cobban, A. (1964). *Rousseau and the Modern State*. London: Archon Books.

Commager, H. S. (1962). *The American Mind*. New Haven: Yale University Press.

—— (1966). *Freedom and Order*. New York: George Braziller.

Comte, A. (1975). *Auguste Comte and Positivism*. (Ed. G. Lenzer). New York: Harper Torch Books.

—— (1988). *Introduction to Positive Philosophy*. Indianapolis: Hackett Publishing.

Condorcet, A-N. de (1955). *Sketch for a Historical Picture of the Progress of the Human Mind* (Trans. J. Barraclough). London: Weidenfeld & Nicolson.

Cooper, D. E. (1996). *Nietzsche*. In N. Bunnin & E. P. Tsui-James. *The Blackwell Companion to Philosophy*. Malden, MA: Blackwell.

Copleston, F. C. (1990, 1972). *A History of Medieval Philosophy*. Notre Dame: University of Notre Dame Press.

—— (1962). *A History of Philosophy*. Vol. *I*, Parts I & II, Garden City, NY: Image Books.

Coppens, Y. (1994). East Side Story: The Origin of Humankind. *Scientific American*, 270 (5), 88-95.

Counts, G. S. (1952). *Education and American Civilization*. New York: Teachers College Press.

Cremin, L. A. (1957). *The Republic and the School, Horace Mann on the Education of Free Men*. New York: Teachers College Press.

—— (1977). *Traditions of American Education*. New York: Basic Books.

Cronbach, L. J. (1963). *Educational psychology*. (2nd ed.). New York: Harcourt Brace.

—— (1990, 1971). *Essentials of Psychological Testing*. (5th ed.). New York: HarperCollins.

—— (1957). "The Two Disciplines of Scientific Psychology." *American Psychologist, 12*, (1957), 671–84.

Crystal, R. G. (1995). "Transfer of Genes to Humans: Early Lessons and Obstacles to Success." *Science*, *270*, 404-410.

Cullen, B. (2003). "Testimony from the Iceman." *Smithsonian*, *33*(11), 42-51.

Damasio, A. R. (1994). *Descartes' Error, Emotion, Reason and the Human Brain*. New York: Avon Books.

—— (1999). *The feeling of what happens, Body and emotion in the making of consciousness.* New York: Harcourt, Brace & Co.

Damrosch, D. (April 1997). "A past we can live with." *Civilization*, 76-77.

Danford, J. W. (1990). *David Hume and the Problem of Reason*. New Haven: Yale University Press.

Darwin, C. (1859, 1979). *The Origin of the Species*. New York: Hill & Wang.

—— (1888). *The Descent of Man*. London: John Murray.

—— (1998). *The Expression of the Emotions in Man and Animals*. London: HarperCollins.

—— (no date, but c.1890). *On the Structure and Distribution of Coral Reefs* (Ed. J. W. Williams). London: Walter Scott Publishing Co.

—— (1979). *The Illustrated Origin of The Species*. New York: Hill &Wang.

—— (1958). *The Autobiography of Charles Darwin and Selected Letters* (Ed. F. Darwin). New York: Dover Publications.

Davis, T, M. & Frost, R. Cohen. (2014). *The Partisan Divide, Congress in Crisis.* Campbell, CA: Premiere.

Dawkins, R. (2004). *The Ancestor's Tale, A Pilgrimage to the Dawn of Life*. Boston: Houghton Mifflin.

Dawson, C. (1958). *Religion and the Rise of Western Culture*. New York: Image Books.

Dennis, C. (2005). "Chimp Genome: Branching Out." *Nature*, *437*, 7055ff.

Descartes, R. (1941, 1637). *A Discourse on Method.* London: J. M. Dent & Sons.

De Tocqueville, A. (1956). *Democracy in America*. New York: New American Library.

Dewey, J. (1964). *Philosophy of Education*. Patterson, NJ: Littlefield, Adams and Co.

—— (1944, 1916). *Democracy and Education*. New York: Free Press

—— (1938, 1958). *Experience and Education*. New York: Dover Publications.

Diamond, J. (2005). *Collapse, How Societies Choose to Fail or Succeed.* New York: Viking.

—— (1997). *Guns, Germs, Steel*. New York: W. W. Norton.

—— (1992). *The Third Chimpanzee*, New York, HarperPerennial.

—— (2012). *The World Until Yesterday, What Can We Learn from Traditional Societies*. New York: Viking.

Dilke, O. A. W. (1987). *Reading the Past, Mathematics and Measurement*. London: The British Museum.

Diogenes Laertius. (1959). *Lives of Eminent Philosophers* (Trans. R. D. Hicks). Vol. 1 & 2. Cambridge: Harvard University Press.

Dreifus, C. (2005, December 6). "Live Longer with Evolution? Evidence May Lie in Fruit Flies." *The New York Times*, D2.

Dudley, W. (Ed.). (2002). *Religion in America, Opposing Viewpoints*. San Diego: Greenhaven Press.

Dundon, R. W. & Jainer, J. L. (1999). *A Philosophy of Social Sciences, An Interdisciplinary Approach*. Commack, NY: Nova Science Publishers.

Durant, W. (1941). *The Mansions of Philosophy*. Garden City, NY: Garden City Publishing Co.

—— (1927). *The Story of Philosophy*. Garden City, NJ: Garden City Publishing Co.

—— (1963, 1935). *Our Oriental Heritage*. New York: Simon & Schuster.

—— (1966). *The Life of Greece*. New York: Simon & Schuster.

—— (1944). *Caesar and Christ*. New York: Simon & Schuster.

—— (1950). *The Age of Faith*. New York: Simon & Schuster.

Durant, W. & Durant, A. (1963). *The Age of Louis XIV*. New York: Simon & Schuster.

—— (1967). *Rousseau and Revolution*. New York: MJF Books.

—— (1975). *The Age of Napoleon*. New York: MJF Books.

—— (1968). *The Lessons of History*. New York: Simon & Schuster.

Durkheim, E. (1938). *The Rules of Sociological Method*. (Trans, S. A. Soloway & J. H. Mueller). New York: The Free Press.

—— (1984). *The Division of Labor in Society*. London: Macmillan.

Ebstein, R. P. (1996). "Dopamine D4 receptor (D4DR) exon III polymorphism associated with the human personality trait of novelty seeking." *Nature Genetics*, *12*, 78-80.

Edgar, B. (September 2014). "Powers of Two." *Scientific American*, Vol. *311*(3), 63-67.

Edmon, I. (1956). *The Works of Plato*. New York: Modern Library.

Ehrlich, P. R. (1968). *The Population Bomb*. New York: Ballantine Books.

Einstein, A. (1945). *The Meaning of Relativity*. Princeton: Princeton University Press.

——— (2009). *A Stubbornly Persistent Illusion, The Essential Scientific Works of Albert Einstein*. (Ed. S. Hawking). Philadelphia: Running Press.

Eliade, M. (1978). *A History of Religious Ideas, from the Stone Age to the Eleusinian Mysteries* (Trans. W. R. Trask) V1. Chicago: University of Chicago Press.

Engels, F. (1985). *The Origin of the Family, Private Property and the State*. Moscow: Progress Publishers.

Erasmus, D. (1887). *The Praise of Folly*. London: Hamilton, Adams.

Erikson, E. H. (1950, 1985). *Childhood and Society*. New York: W.W. Norton.

—— (1968). *Identity, Youth and Crisis*. New York: W.W. Norton.

Erwin, D. H. (1996, July). "The Mother of Mass Extinctions." *Scientific American*, 72-78.

Etzioni, A. (April, 1961). "Review: Max Weber as an Intellectual." *American Journal of Economics and Sociology.*, *20*(3), 331–333.

Evans, R. J. (1997). *In Defence of History*. London: Granta.

Evans, J. A. S. (1982). *Herodotus*. Boston: Twayne Publishers.

Eysenck, M. W. (1993). *Principles of Cognitive Psychology*. Hove, UK: Lawrence Erblaum.

Eysenck, H. J. (1967). "Intelligence assessment: a theoretical and experimental approach." *British Journal of Educational Psychology*, *37*, 81-98.

—— (1973). *The Measurement of Intelligence*. Baltimore: The Williams & Wilkins Co.

Fagan, B. M. (1990). *The Journey from Eden, the Peopling of our World*. London: Thames & Hudson.

—— (2013). *The Attacking Ocean, The Past, Present and Future of Rising Sea Levels*. New York: Bloomsbury.

Ferguson, N. (2012). *Civilization*. London: Penguin Books.

Fernandez-Armesto, F. (2000). *Civilizations*. London: Macmillan.

Festinger, L. (1957). *A Theory of Cognitive Dissonance*. Stanford, CA: Stanford University Press.

Flexner, J. T. (1978). *The Young Hamilton, a Biography*. Boston: Little, Brown.

Finlayson, C. (2014). *The Improbable Primate, How Water Shaped Human Evolution.* Oxford, UK: Oxford University Press.

Foer, J. (May 2015). "Understanding Dolphins, Intelligence, Captivity, Culture." *National Geographic*, *227*(5), 35-54.

Foley, D. (2006). *Adam's Fallacy, A Guide to Economic Theology*. Cambridge, MA: The Belknap Press of Harvard University Press.

Foroohar, R., "Wall Street's Values are Strangling American Business," *Time*, July 21, 2014, p. 14.

Foucault, M. (1972). *The Archeology of Knowledge*. (Trans. A. L. Sheridan Smith). New York: Pantheon.

Frame, D. M. (1955). *Montaigne's Discovery of Man, the Humanization of a Humanist.* New York: Columbia University Press.

Franklin, B. (1964). *The Autobiography of Benjamin Franklin.* New Haven: Yale University Press.

—— (1970). *The Complete Poor Richard's Almanacs.* (Vol. 1) Barre, MA: Imprint Society.

France, P. (1993). *Greek as a Treat, an Introduction to the Classics.* London: BBC Books.

Freeman, D. S. (1968, 1948). *Washington* (Abridged by R. Harwell), New York: Charles Scribner's Sons.

Freeman, C. (2003). *The Closing of the Western Mind, The Rise of Faith and The Fall of Reason.* New York: Knopf.

Freud, S. (1952). *The Ego and the Id.* In Adler, M. J., (Ed.), *Great Books of The Western World, V.* 54, Chicago: Encyclopaedia Britannia, 697-717.

—— (1910, 1962). *Three Contributions to the Theory of Sex*. (Trans. by A. A. Brill) New York: E. P. Dutton.

Friedman, R. A. (June 29, 2014). "Why Teenagers Act Crazy." *The New York Times*, sr1ff.

Fukuyama, F. (September/October 2014). "America in Decay, The Sources of Political Dysfunction." *Foreign Affairs*, *93*(5), 5-26.

Gall, C. (2014). *The Wrong Enemy, America in Afghanistan 2001-2014.* New York: Penguin Books.

Galton, F. (1978). *Hereditary Genius, An Inquiry into its Laws and Consequences*. New York: St. Martin's Press.

Gassendo, P. (1647). *Institutio astronomica juxta hypotheseis tam veterum, quam copernici et tychonis*. Paris: Ludovicum de Hevqueville.

Gay, P. (1988). *Freud, a Life for our Time*. New York: Norton.

Gazzaniga, M. S. (1992). *Nature's Mind, the Biological Roots of Thinking, Emotions, Sexuality, Language and Intelligence*. New York: Basic Books.

—— (1995). *The Cognitive Neurosciences*. Cambridge: The MIT Press.

Gaukroger, S. (1995). *Descartes, an Intellectual Biography*. Oxford: The Clarendon Press.

Gaur, A. (1984). *A History of Writing*. New York: Charles Scribner's Sons.

Giancola, P, & Zeichner, A. (1994). "Neurophysiological performance on tests of frontal-lobe functioning and aggressive behavior in men." *J. of Abnormal Psychology*, *103*(4), 832-835.

Gibbon, E. (1985, 1952). *The Decline and Fall of the Roman Empire*. New York: Penguin Books.

Gilgamesh (1960). *The Epic of Gilgamesh* (Trans. N. K. Sandars), Hammondsworth, England: Penguin Books.

Gill, J. B. (1984). *The Great Pyramid Speaks, an Adventure in Mathematical Archaeology*. New York: Barnes & Noble.

Gilson, E. (1991, 1936). *The Spirit of Medieval Philosophy*. New York: Charles Scribner's.

Ginsberg, E. (1934). *The House of Adam Smith*. New York: Columbia University Press.

Gopnik, A. (October 2015). "David Hume and the Buddha." *The Atlantic*, 96-110.

Gore, R. (2002, August). "The First Pioneer, A New Find Shakes the Human Tree." *National Geographic*, special edition.

—— (2003). *The Rise of Mammals, Adapting, Evolving, Surviving*.

Gottschall, J. (2015). *The Professor in the Cage, Why Men Fight and Why We Like to Watch*. New York: Penguin Press.

Gould, S. J. (1989). *Wonderful Life, the Burgess Shale and the Nature of History*. New York: W.W. Norton.

—— (1994, October). "The Evolution of Life on Earth." *Scientific American*, 85-91.

—— (2002). *The Structure of Evolutionary Theory*. Cambridge: Harvard University Press.

Grandin, T. (2012). *The Autistic Brain, Helping Different Kinds of Minds Succeed*. Boston, Mariner Books.

Grant, M. (1992). *Readings in the Classical Historians*. New York: Charles Scribner's Sons.

Gutek, G. L. (1972). *A History of the Western Educational Experience*. New York: Random House.

Guthrie, W. K. C. (1975,1950). *The Greek Philosophers, from Thales to Aristotle*. New York: Harper Torchbooks.

Habashi, F. (1998). "The Age of Alchemy: History of Chemistry, Metallurgy And Civilisation." *Interdisciplinary Science Reviews*, *28* (4), 348-61.

Habermas, J. (1987). *The Philosophical Discourse on Modernity*. Cambridge: The MIT Press.

Hacking, I. (1975). *Why does Language Matter to Philosophy*? Cambridge: Cambridge University Press.

Hagel, C. (2014). *America, Our Next Chapter*. New York: Ecco Press.

Hainds, J. R. (1946, January). "John Stuart Mill and the Saint Simonians." *Journal of the History of Ideas*, *7*(1), 103–112.

Hall, E.T. (1959). *The Silent Language*. New York: Doubleday.

Hadas, M. (1951). *The Basic Works of Cicero*. New York: The Modern Library.

Hakluyt, R. (1589). *The Principal Navigations, Voyages and Discoveries of The English Nation*. London: George Bishop & Ralph Newberie.

Hale, J. (1994). *The Civilization of Europe in the Renaissance*. New York: Atheneum.

Hamilton, E. (1993, 1930). *The Greek Way*. New York: W. W. Norton.

—— (1957). *The Echo of Greece*. New York: W. W. Norton.

—— (1964, 1932). *The Roman Way*. New York: W. W. Norton.

—— (1948). *Witness to the Truth, Christ and His Interpreters*. New York: W. W. Norton.

Harlow, H. F. & Mears, C. (1979). *The Human Model: Primate Perspectives*. New York: John Wiley.

Harrison, J. & Laslett, P. (1971). *The Library of John Locke*. Oxford: The Clarendon Press.

Hartman, E. (1991). *Boundaries in the Mind, a New Psychology of Personality*. New York: Basic Books.

Hawking, S. W. (1988). *A Brief History of Time*. New York: Bantam Books.

Healey, J. F. (1990). *Reading the Past, the Early Alphabet*. London: The British Museum.

Heilbroner, R. L. (1972). *The Worldly Philosophers, the Lives, Times and Ideas of the Great Economic Thinkers* (4th ed.). New York: Touchstone Book.
Heider, F. (1958). *The Psychology of Interpersonal Relations*. New York: Wiley.
Hegel, G. W. F. (1888). *Lectures on the Philosophy of History*. London: George Bell & Sons.
Heilbroner, R. & L. Thurow. (1998). *Economics Explained, Everything You Need to Know About How the Economy Works and Where It's Going*. New York: Touchstone Books.
Herodotus (1830). *The Histories*. (v.1) London: Henry Colburn & Richard Bentley.
Heun, M. (1997). "Site of einkorn wheat domestication identified by DNA fingerprinting." *Science*, *278*, 1312-14.
Highet, G. (1954). *Man's Unconquerable Mind*. New York: Columbia University Press.
Hilgard, E. C. (1948, 1966). *Theories of Learning*. (3rd ed.) New York: Appleton-Century-Crofts.
Hirsch, E. D. (1987). *Cultural Literacy: What Every American Needs to Know*. Boston: Houghton Mifflin.
—— (1993). *The Dictionary of Cultural Literacy*. Boston: Houghton Mifflin.
Hobbes, T. (1651). *Leviathan*. London: J. M. Dent & Sons.
—— *English Works of Thomas Hobbes* (1839, 1962). London: Scientia Aalen.
Hodder, I. (1986). *Reading the Past, Current Approaches to Interpretation in Archaeology*. Cambridge: Cambridge University Press.
Hofstadter, R. (1989, 1948). *The American Political Tradition, And the Men Who Made it*. New York: Vintage Books.
Holland, E. (January 2015). "When Dinosaurs Were Doomed," *Smithsonian*. Vol. *45*(9), 63-66.
Holland, J. H. (1998). *Adaptation in Natural and Artificial Systems, An Introductory Analysis with Applications to Biology, Control and Artificial Intelligence*. Cambridge: The MIT Press.
Hollis, K. (May 2015). "Behavior and Learning are Key Elements in Evolutionary Arms Race." APA Science Brief retrieved at: http:// www.apa.org/ science/ about/ psa/ 2015/ 05/evolutionary-arms.aspx.
Hollis, M. (1996). Philosophy of Social Science. In N, Bunnin & E. P. Tsui-James, *The Blackwell Companion to Philosophy*. Malden, MA: Blackwell.
Holt, J. (2012). *Why Does the World Exist?* New York: Liveright Publications.
Homer (1974). *The Iliad*. Garden City, NJ: Doubleday.
—— (1974). *The Odyssey*. Garden City, NJ: Doubleday.
Hozeski, B. (Tr.). (1994). *Hildegard of Bingen*. New York: Oxford University Press.
Huizinga, J. (1996, 1921). *The Autumn of the Middle Ages*. (R. J. Payton, U. Mammitzsch, trans.) Chicago: University of Chicago Press.
—— (1959). *Men and Ideas, History, the Middle Ages, the Renaissance*. New York: Meridian Books.
Hume, D. (1967a). *On Human Nature and the Understanding*. New York: Collier Books.
—— (1967b). *A Treatise of Human Nature*. Oxford: The Clarendon Press.
—— (1779, 1969). *Dialogues Concerning Natural Religion*. New York: Hafner Publishing Co.,
Hundert, E.M. (1989). *Philosophy, Psychiatry and Neuroscience, Three Approaches to the Mind*. Oxford: Clarendon Press.

Huntington, S. P. (1996). *The Clash of Civilizations and the Remaking of World Order.* Cambridge: Harvard University Press.

International Monetary Fund (IMF). (2014). *Fiscal policy and income inequality.* Retrieved from http://www.imf.org/external/np/fad/inequality/

Isaacs, H. (1972). *Idols of the Tribe, Group Identity and Political Change.* Cambridge: Harvard University Press.

James, W. (1892). *Psychology.* New York: Henry Holt and Company.

—— (1997). *The Varieties of Religious Experience.* New York: Touchstone Books.

Jefferson, T. (1972). *Notes on the State of Virginia.* New York: W. W. Norton. & Co.

Jenkins, K. (1991). *Re-thinking History.* London: Routledge.

Jensen, A. R. (1969). "How much can we boost IQ and scholastic achievement?" *Harvard Educational Review, 39,* 1-123.

—— (1980). *Bias in Mental Testing.* New York: Harper and Row.

—— (1972). *Genetics and Intelligence.* New York: Harper & Row.

Johanson, D. (1996). "Human origins." *National Forum, 76* (1), 24-27.

Johanson, D. & Edey, M. (1981). *Lucy, the Beginnings of Humankind.* New York: Simon & Schuster.

Johanson, D. & Johanson, L. (1994). *Ancestors, In Search of Human Origins.* New York: Villard Books.

Jones, J. (1999). *A Social History of the Laboring Classes, from Colonial Times to the present.* Oxford, UK: Blackwell.

Jones, S. (1993). *The Language of the Genes, Solving the Mysteries of our Genetic Past, Present and Future.* New York: Anchor Books.

—— (1996). *In the Blood: God, Genes and Destiny.* London: HarperCollins.

Josephson, M. (1934). *The Robber Barons, The Great American Capitalists, 1861-1901.* New York: Harcourt, Brace and Co.

Jung, C. G. (1954). *The Development of Personality.* Princeton: Princeton University Press.

—— (1958, 1957). *The Undiscovered Self.* New York: Penguin

—— (1966). *Two Essays on analytical psychology*, Vol. *7*. Princeton: Princeton University Press.

—— (1969, 1959). *The Archetypes and the Collective Unconscious*, V. *9* Part 1. Princeton: Princeton University Press.

—— (1971). *The Portable Jung.* New York: Penguin Books.

—— (1979). *Word and Image.* Princeton: Princeton University Press.

Junger, S. (June 2015). "The Bonds of Battle." *Vanity Fair, 658,* 106-111.

Kagan, J. (1994). *Galen's Prophecy, Temperament in Human Nature.* New York: Basic Books.

Kahn, P. & Gibbons, A. (1997). "DNA from an extinct human." *Science, 277,* 176-78.

Kant, I. (1949). *Critique of Pure Reason.* Chicago: University of Chicago Press.

Kaplan, F. (2014). *John Quincy Adams, American Visionary.* New York: Harper.

Kennedy, J. F. (1964). *Profiles in Courage.* New York: Harper & Row.

Keynes, J. M. (1973). *The Collected Writings of John Maynard Keynes.* Vol. VII, *The General Theory of Employment, Interest and Money* Cambridge, UK: Macmillan/St. Martin's Press.

Kirk, G. S., Raven, J. E. & Schofield, M. (1983, 1957). *The Presocratic Philosophers.* Cambridge, UK: Cambridge University Press.

Klosko, G. (1986). *The Development of Plato's Political Theory*. New York: Metheun.

Knight, H. (May 17, 2015) “Racial Islands in San Francisco’s Sea of Diversity,” *San Francisco Chronicle*, Section R.

Koch, A. (1973, 1950). *Jefferson and Madison, the Great Collaboration*. London: Oxford University Press.

Kohler, W. (1925). *The Mentality of Apes*. New York: Harcourt, Brace & Co.

—— (1969). *The Task of Gestalt Psychology*. Princeton: Princeton University Press.

Kolbert, E. (2014). *The Sixth Extinction, An Unnatural History*. New York: Henry Holt & Co.

—— (August 31, 2015). “The Terrible Teens.” *The New Yorker*, 83-86.

Kosslyn, S.M. & Koenig, O. (1992). *Wet Mind, the New Cognitive Neuroscience*. New York: The Free Press.

Krauss, L. M., “A Beacon from the Big Bang,” *Scientific American*, *311*(4), 59-67.

Kuhn, T. S. (1985, 1957). *The Copernican Revolution, Planetary astronomy in the Development of Western Thought*. Cambridge: Harvard University Press.

—— (1970, 1962). *The Structure of Scientific Revolutions*. Chicago: University of Chicago Press.

Lavine, A. (1984). *From Socrates to Sartre*. New York: Bantam Books.

Lazaridis, I, *et al.*, “Ancient Human Genomes Suggest Three Ancestral Populations for Present-day Europeans,” (September 18, 2014) *Nature*, *513*, 409-413.

Leakey, R. & Lewin, R. (1992). *Origins reconsidered, in search of What Makes Us Human*. New York: Doubleday.

Learner, R. E. (1988). *Western civilizations, their history and their culture*. (11th Ed.). New York: W. W. Norton.

Levy, L. (Ed.) (1969). *Essays on the Making of the Constitution*. New York: Oxford University Press.

Lewin, K. (1936). *Principles of Topological Psychology*. New York: McGraw-Hill.

Lieberman, D. E. (2007, September 20). “Palaeoanthropology: Homing in on early Homo.” *Nature*, *449*, 291–292.

Lipton, E. (December 7, 2014). “Energy and Regulators on One Team.” *The New York Times*, A1, 30-31.

Locke, J. (1714). *The Works of John Locke*. London: Churchill.

—— (1961). *An Essay Concerning Human Understanding*. New York: E. P. Putnam.

—— (1964). *Some Thoughts Concerning Education*. London: Heineman.

Lorenz, K. (1963). *On Aggression*. New York: Harcourt, Brace & World.

Lukianff G. & J. Haidt, “The Coddling of the American Mind, *The Atlantic Monthly*, *316*(2), 42-52.

Lucretius, (1952). *On the Nature of Things*. In R. M. Hutchins (Ed.). *Great Books of the Western World*. Chicago: Encyclopaedia Britannia.

Luders, E., C. Gaser, K. L. Narr, & A. W. Toga (November 11, 2009). “Why Sex Matters: Brain Size Independent Difference in Gray Matter Distributions Between Men and Women,” *J. of Neuroscience*, *29*(45), 14265-14270.

Luhrmann, T. H., “The Anxious Americans,” *The New York Times*, July 19, 2015, p. Sr. 9.

Lustgarten, A. “How the West Overcounts Its Water,” *The New York Times*, July 19, 2015, p. Sr. 1 & Sr. 6.

Machiavelli, N. (1999). *The Prince*. London: Penguin Books.

Maddow, R. (2012). *Drift, The Unmooring of American Military Power*, New York: Crown Publishers.

Madison, J. (1961). *The Federalist Papers*. New York: New American Library.

—— (1893). *Journal of the Constitutional Convention*. Chicago: Scott, Foresman & Co.

Magill, F. N. (1990, 1961). *Masterpieces of World Philosophy*. New York: HarperCollins.

Magnusson, M. (1992). *Vikings*! London: BBC Books.

Mall, R. (1967). *Hume's Concept of Man, an Essay in Philosophical Anthropology*. Bombay: Allied Publishers.

Malthus, T. R. (1963, 1798). *An Essay on the Principle of Population*. Homewood, ILL: Richard D. Irwin, Inc.

Maurois, A. (1943, 1932). *Voltaire*. London: Thomas Nelson & Sons.

Maziarz, E. A. & Greenwood, T. (1968). *Greek Mathematical Philosophy*. New York: Barnes & Noble.

McConica, J., Quinton, A., Kenny, A. & Burke, P. (1993). *Renaissance Thinkers*. New York: Oxford University Press.

—— (1991). *Erasmus*. Oxford: Oxford University Press.

McDonald, F. (1985). *Novus Ordo Seculorum, The Intellectual Origins of the Constitution*. Lawrence, KS: University Press of Kansas.

McGrath, A. (1987). *The Intellectual Origins of the European Reformation*. Oxford, England: Blackwell.

McKeon, R. (1947). *Introduction to Aristotle*. New York: Modern Library.

Mead, M. (1949,1928). *Coming of Age in Samoa*. New York: The New American Library.

Mellor, R. (1993). *Tacitus*. New York: Routledge.

Menand, L. (2001). *The Metaphysical Club*. New York: Farrar, Straus, & Giroux.

Mencken, H. L. (1948). *Treatise on the Gods*. New York: Alfred A. Knopf.

Merleau-Ponty, M. (1966, 1942). *Phenomenology of Perception*. London: Routledge & Kegan Paul.

Mill, J. S. (1939). "Utilitarianism." In Burtt, E.A. *The English Philosophers from Bacon to Mill*. New York: The Modern Library, 895-948.

—— (1969). *Autobiography*. (Ed. J. Stillinger). Boston: Houghton Mifflin.

—— (1977). *Essays on Politics and Society*. Toronto: University of Toronto Press.

—— (1965). *Principles of Political Economy, with Some of Their Applications to Social Philosophy*. Toronto: University of Toronto Press.

—— (1994). *Principles of Political and Economy and Chapters on Socialism*. New York: Oxford University Press.

Miller, E. L. (1992). *Questions that Matter, an Invitation to Philosophy*. New York: McGraw Hill.

Montaigne, M. (1957). *The Complete Works of Montaigne, Essays, Travel, Journal, Letters*. (Tr: D. M. Frame) Stanford: Stanford University Press.

Montessori, M. (1913). *Pedagogical Anthropology*. New York: J. P. Lippincott.

Moorey, P. R. S. (1994). *The Ancient Near East*. Oxford: The Ashmolean Museum.

More, T. (1949, 1516). *Utopia*. New York: Appleton-Century-Crofts.

Morris, C. R. (1959). *Locke, Berkeley, Hume*. Oxford: Oxford University Press.

—— (2005). *The Tycoons*. New York: Henry Holt & Co.

Moser, B. (July 6, 2014). "Patriarch and Pariah." *The New York Times Book Review*, p. 17.

Moyers, B. (1990). *A World of Ideas II*. New York: Doubleday.

Muller, J. Z. (March/April 2013). "Capitalism and Inequality." *Foreign Affairs*, *92*(2), 30-51.

Murdoch, I. (1999). *Existentialists and Mystics, Writings on Philosophy and Literature*. New York: Penguin Books.

Murray, C. (1994). *The Bell Curve, Intelligence and Class Structure in American Life*. New York: The Free Press.

—— (2003). *Human Accomplishment, The Pursuit of Excellence in the Arts and Sciences, 800 B.C. to 1950*. New York: Perennial.

Myrdal, G. (1962, 1944). *An American Dilemma, The Negro Problem and Modern Democracy*. New York: Harper & Row.

Naisbitt, J. (1982). *Megatrends*. New York: Warner Books.

Naisbitt, J. & Aburdene, P. (1990). *Megatrends 2000*. New York: William Morrow & Co.

Nasr, V. (2013). *The Indispensable Nation, American Foreign Policy in Retreat*. New York: Anchor Books.

Nehamas, A. (1997). "Trends in Recent American Philosophy." *Daedalus*, *126* (4).

Newman, J. H. (1947). *The Idea of a University*. New York: Longmans, Green & Co.

Nye, B. (2014). *Undeniable, Evolution and the Science of Creation*. New York: St. Martin's Press.

Nietzsche, F. (2002). *Beyond Good and Evil*. Cambridge: Cambridge University Press.

—— (1966). *Thus Spake Zarathustra, A Book for All and None*. (Trans. W. Kaufmann). New York: Viking Penguin.

—— (1990). *Twilight of the Idols and The Anti-Christ*. New York: Penguin Books.

Nijhuis, M. (October 2014). "When the Snows Fail." *National Geographic*, *226*(4), 56-77.

Norwich, J. J. (1998). *A Short History of Byzantium*. London: Penguin Books.

Nye, B. (2014). *Undeniable, Evolution and the Science of Creation*. New York: St. Martin's Press.

Organisation for Economic Cooperation and Development (OECD). (2008). *Economic Survey of the United States*. Retrieved From: http://www.oecd.org/unitedstates.

O'Grady, J. (1985). *Early Christian Heresies*. New York: Barnes & Noble.

Olsen, R. G. (2004). *Science and Religion, 1450-1900, From Copernicus to Darwin*. Baltimore: The Johns Hopkins University Press.

Ogg, O. (1983). *The 26 Letters*. New York: Van Nostrand.

Orgel, L.E. (1994). "The Origin of Life on the Earth." *Scientific American*, *271*(4), 77-83.

Ornstein, R. (1993). *The Roots of the Self, Unraveling the Mystery of Who We Are*. San Francisco: Harper.

Paabo, S. (1993). "Ancient DNA." *Scientific American*, *269* (5), 86-92.

Packer, G. (2013). *The Unwinding, An Inner History of the New America*. New York: Farrar, Straus and Giroux.

Padover, S. K. (1967). *The Washington Papers*. New York: Grosset & Dunlap.

Paine, T. (1993). *The Age of Reason*. New York: Gramercy Books.

Panetta, L. (2014). *Worthy Fights*. New York: Penguin Press.

Pater, W. (1986, 1873). *The Renaissance*. London: Penguin Books.

Pavlov, I. P. (1927). *Conditioned Reflexes, an Investigation of the Physiological Activity of the Cerebral Cortex*. Oxford: Oxford University Press.

—— (1966). *Essential Works of Pavlov*. New York: Bantam Books.

Patti, T. S. (1995). *Manuscripts of the Bible*. London: The British Library.

Perry, L. (1989). *Intellectual Life in America, a History*. Chicago: University of Chicago Press.

Pestalozzi, H. (1951). *The Education of Man*. New York: Greenwood.

Phillips, K. (2008). *Bad Money, Reckless Finance, Failed Politics, and the Global Crisis of American Capitalism*. New York: Penguin Books.

Piketty, T. (2014). *Capital in the twenty-first century*. Cambridge, MA: Belnap Press.

Pinker, S. (1994). *The language Instinct, How the Mind Creates Language*. New York: HarperPerennial.

—— (2002). *The Blank State, The Modern Denial of Human Nature*. New York: Viking.

PISA Report (2009). Paris: Organisation of Economic Cooperation and Development.

Plato (1944). *Republic*. (tr. B. Jowett) New York: The Heritage Press.

—— (1961). *The Collected Dialogues*. Princeton: Princeton University Press.

—— (1994). *Symposium*. (Tr. R. Waterfield). Oxford: Oxford University Press.

Plomin, R. (1986). *Development, genetics and psychology*. Hillsdale, NJ: Lawrence Erlbaum Associates, Publishers.

—— (1994). *Genetics and experience*. Thousand Oaks, CA: Sage Publications.

Plutarch (1967). *The Lives of the Noble Grecians and Romans*. New York: Modern Library.

Polo, M. (1958). *The Travels of Marco Polo*. (trans. R. Latham). London: Penguin Books.

Porter, E. (2014, March 13). "A relentless widening of disparity in wealth." *The New York Times*, A1.

Powicke, M. (1953). *The Thirteenth Century, 1216-1307*. Oxford, UK: The Clarendon Press.

Preston, D. (September 2014). "The 9,000-year-old Man Speaks." *Smithsonian*, Vol. 46(5), pp. 52-63.

Proust, M. (1934). *Remembrance of Things Past*. (Vol. 2). New York: Random House.

Rabb, T. K. (1993). *Renaissance Lives, Portraits of an Age*. New York: Pantheon.

Raleigh, W. (1708). *The General History of the World*. London: Bell, Smith & Round.

Rayner, B. L. (1834). *Life of Thomas Jefferson*. Boston: Tilly, Wait, Colman & Holden.

"Reefs at Risk." (June 2015). *National Geographic*, Vol. *227*(6).

Report on the Economic Well-Being of U.S. Households in 2014. Washington, DC: Board of Governors of the Federal Reserve System.

Reston, J. (1994). *Galileo, a Life*. New York: HarperPerennial.

Ricard, M, A. Lutz, & R. J. Davidson, "Mind of the Meditator." *Scientific American*, November 2014, *311*(5), 39-45.

"Rising Temperatures on Earth Could Threaten Human Life Congress Told." *The Houston Post*, March 3, 1982.

Roberts, C. (2012). *The Ocean of Life, The Fate of Man and the Sea*. New York: Penguin Books.

Roberts, J. M. (1993, 1976). *History of the World*. New York: Oxford University Press.

Roberts, P. (2014). *The Impulse Society. America in the Age of Instant Gratification*. New York: Bloomsbury.

Robson, J. M. & Stillinger, J. (Eds.). (1981). *Collected Works of John Stuart Mill*. Toronto: University of Toronto Press.

Rouse, W. H. D. (1956). *Great Dialogues of Plato*. New York: New American Library.

Rossiter, C. (Ed.) (1961). *The Federalist Papers*. New York: New American Library.

Ross, I. S. (1995). *The Life of Adam Smith*. Oxford, UK: Clarendon Press.

Rousseau, J. J. (1964, 1768). *Emile, Julie and Other Writings*.

Woodbury, NY: Barrons Educational Series, Inc.
——— (1968). *The Social Contract*. New York: Penguin Classics.
Russell, B. (1955). *A History of Western Philosophy*. London: George Allen & Unwin.
—— (1967, 1932). *Education and the Social Order*. London: Unwin.
—— (1961). *The Analysis of Mind*. London: Geroge Allen & Unwin.
—— (1948). *Human Knowledge, its Scope and Limits*. New York: Simon and Schuster.
—— (1997, 1935). *Religion and Science*. New York: Oxford University Press.
Sachs, J. D. (June 2008). "Surging Food Prices and Global Stability." *Scientific American, 298*(6), 40.
Sagan, C. (1977). *The Dragons of Eden, Speculations on the Origin of Human Intelligence*. New York: Random House.
Sandel, M. J. (1996). *Democracy's Discontent, America in Search of a Public Philosophy*. Cambridge: Harvard University Press.
Sapir, E. (1921, 1961). *Culture, Language and Personality*. Berkeley: University of California Press.
—— (1949, 1921). *Language, an Introduction to the Study of Speech*. New York: Harcourt, Brace & Co.
Sapolsky, R. M. (2001). *A Primate's Memoir, A Neuroscientist's Unconventional Life Among the Baboons*. New York: Touchstone Books.
Sartre, J. P. (1956). *Being and Nothingness*. New York: Philosophical Library.
—— (1964). *Nausea*. New York: New Directions Press.
—— (1972). *The Psychology of Imagination*. London: Methuen.
Schachner, N. (1957). *Thomas Jefferson*. New York: Thomas Yoseloff.
Scharf, C. (August 2014). "Cosmic (In)Significance," *Scientific American, 311*(2), 74-77.
Schlesinger, A. M., Jr. (1958). *The Coming of the New Deal*. Boston: Houghton Mifflin Co.
Schopenhauer, A. (1998). *The World as Will and Idea*. London: J. M. Dent.
Scriven, M. (1966). *Primary Philosophy*. New York: McGraw Hill.
Seneca (1997). *Dialogues and Letters* (C. D. N. Costa, Ed. & Trans.) London: Penguin Books.
Sharpes, D. (2002). *Advanced Educational Foundations, The History, Philosophy and Culture of Schools*. New York: Routledge.
——(1999). *Advanced Psychology for Teachers*. New York: McGraw Hill.
——(1988). *Education and the U.S. Government*. New York: St. Martin's Press.
——(2012). *Our Divided House, Anti-Government Movements and the Failure of Unification*. Vulcan: CreateSpace, a Division of Amazon.
——(August 2014). "American Economic Decline: A Consequence of Lower Educational Standards." *US-China Education Review*, Vol. *4*(8), 560-66.
Sharpes, D. K. & Wang, X. (1997). "Adolescent Self-Concept among Han, Mongolian and Korean Chinese." *Adolescence, 32* (128), 913-924.
Shepherd, G. M. (1994). *Neurobiology. (3rd Ed.)*. New York: Oxford University Press.
Sherwin, M. J. (Fall 2012). "One Step from Nuclear War: The Cuban Missile Crisis at 50." *Quarterly of the National Archives*, Vol. *44*(3), 6-16.
Sinclair, U. (2012, 1906). *The Jungle*. New York: Barnes and Noble.
Skinner, B. F. (1938). *The Behavior of Organisms, an Experimental Analysis*. New York: Appleton-Century-Crofts.
—— (1948). *Walden Two*. New York: Macmillan.

—— (1953). *Science and human behavior*. New York: Macmillan.
—— (1968). *The Technology of Teaching*. New York: Appleton-Century-Crofts.
—— (1971). *Beyond Freedom and Dignity*. New York: Bantam Books.
—— (1974). *About Behaviorism*. New York: Knopf.
—— (1976). *Particulars of my Life*. New York: Alfred Knopf
—— (1978). *Reflections on Behaviorism and Society*. Englewood Cliffs, NJ: Prentice-Hall.
—— (1979). *The Shaping of a Behaviorist*. New York: Alfred Knopf.
—— (1984). *A Matter of Consequences*. New York: New York University Press.
Skinner, M. K. (August 2014). "A New Kind of Inheritance," *Scientific American*, Vol. *311*(2), 45-51.
Schlesinger, A. M. Jr. (1958). *The Coming of the New Deal*. Boston: Houghton Mifflin Co.
Smith, A. (1991). *The Wealth of Nations*. Amherst, NY: Prometheus Books.
Solecki, R. S. (1971). *Shanidar, the First Flower People*. New York: Alfred Knopf.
Sophocles. (1977). *The Oedipus Cycle*. (trans. D. Fitts & R. Fitzgerald). New York: Harvest Book, Harcourt, Inc.
Sorell, T. (1996). *The Cambridge Companion to Hobbes*. Cambridge: Cambridge University Press.
Spencer, H. (1906, 1861). *Education, Intellectual, Moral and Physical*. London: Watts & Co.
—— (1870). *First Principles*. London: Williams and Norgate.
Spindler, K. (1994). *The Man in the Ice*. New York: Harmony Books.
Spinoza, B. (1894). *Ethics*. London: T. Fisher Unwin.
Sulloway, F. J. (1996). *Born to Rebel: Birth Order, Family Dynamics, and Creative Lives*. New York: Pantheon Books.
Surowiecki, J. (September 24, 2015). "Why the Rich Are So Much Richer." *The New York Review of Books*, LXII, *14*, pp. 32-36.
Stanford Center on Poverty and Inequality. (2014). Retrieved from http:// web.stanford.edu /group/scspi/ *Statistical Programs of the United States Government*. (2014). Washington, DC: Office of Management and Budget.
Stiglitz, J. (2013). *The Price of Inequality*. New York: W. W. Norton.
Stowe, H. B. (1882). *Uncle Tom's Cabin*. Chicago: M.A. Donohue & Co.
Stroope, S., McFarland, M. J. & Uecker, J. E. (2015). "Marital characteristics and the Sexual relationships of U.S. older adults." *Archives of Sexual Behavior*. *44*, (1), 233-247.
Stroud, D. I. (1983). *Magna Carta*. Ringwood, UK: Brown & Sons.
Stryker, N. (2014). *The Thing with Feathers, The Surprising Lives of Birds And What they Reveal About Being Human*. New York: Riverhead Books.
Synnott, M. (June 2015). "Sins of the Aral Sea." *National Geographic*, Vol. *227*(6), 114-29.
Tactitus (tr. K. Wellesley). (1995). *The Histories*. London: Penguin Books.
Tagore, R. (1988). *The Religion of Man*. London: Unwin Paperbacks.
Tarnas, R. (1991). *The Passion of the Western Mind, Understanding the Ideas That Have Shaped our World View*. New York: Ballantine.
Tattersall, I. (2001). "How We Came to be Human." *Scientific American*, *285*(6), 57-63.
Tawney, R. H. (no date but 1926). *The Acquisitive Society*. New York: Harcourt, Brace & Co.
Taylor, P. (1960). *The Notebooks of Leonardo da Vinci*. New York: New American Library.
Thomas, L. (1974). *The Lives of a Cell*. New York: Bantam Books.
Thompson, D. (July/August 2015). "A World Without Work," *The Atlantic*, *316*(1), 51-61.
Thorndike, E. L. (1911). *Individuality*. Boston: Houghton Mifflin.

—— (1914, 1923). *Educational Psychology, Briefer Course*. New York: Teachers College.

—— (1931). *Human Learning*. Cambridge, MA: MIT Press.

—— (1913). *The Original Nature of Man, Educational Psychology V.1*. New York: Teachers College.

Thucyidides (Tr. R. Warner). *History of the Peloponnesian War*. London: Penguin Books.

Tingley, K. (July 6, 2014) "One Child, Three Parents." *The New York Times Magazine*, 26ff.

Tomasello, M. (May 25, 2008). "How Are Humans Unique?" *The New York Times Magazine*, p.15.

Toulmin, S. (1990). *Cosmopolis, the Hidden Agenda of Modernity*. New York: The Free Press.

Toynbee, A. J. (1946). *A Study of History*. London: Oxford University Press.

Trinkaus, E. (2003). "An Early Modern Human from the Petera cu Oase, Romania." *Proceedings of the National Academy of Sciences*, *100*, 11231–11236.

Tuck, R. (1989) *Hobbes*. Oxford: Oxford University Press.

Tudge, C. & J. Young. (2009). *The Link, Uncovering Our Earliest Ancestor*. New York: Little, Brown and Co.

United Nations Development Programme. (2013). *2013 Human Development Report*. Retrieved from http://hdr.undp.org/en/2013-report.

Van Doren, C. (1991). *A History of Knowledge, the Pivotal Events, People, And Achievements of World History*. New York: Ballantine Books.

Veblen, T. (1924). *Theory of the Leisure Class*. New York: Viking Press.

Vernant, J-P. (1982) *The Origins of Greek Thought*. New York: Cornell University Press.

Vico, G. (1725, 1948). *The New Science of Giambattista Vico*. (Ed. & Trans. from 3rd Ed., 1744, by Bergin, T. G., Fisch, M. H). Ithaca, NY: Cornell University Press.

Virgil (1983). *The Aeneid*. New York: Random House.

Voltaire (1972). *Philosophical Dictionary*. (Ed., T. Besterman). New York: Penguin Books.

—— (1956). *Candide and Other Writings*. New York: The Modern Library.

Vygotsky, L S. (1962, 1978). *Mind in society: the development of the higher psychological processes*. (M. Cole, V. John-Steiner, S. Scribner, & E. Souberman, Eds.). Cambridge: Harvard University Press.

—— (1986). *Thought and Language*. Cambridge: MIT Press.

Wade, N. (May 5, 2015). A New Order to Life's Origins. *The New York Times*, D4.

Wallace, A. R. (1902). *The Malay Archipelago, The Land of the Orang-Utan and the Bird of Paradise*. London: Macmillan and Co.

Wallbank, T. W. (1992). *Civilization, Past and Present*. New York HarperCollins.

Walker, D. (2014). *Quarterly Update: Foreign Ownership of US Assets*. Retrieved from http://www.cfr.org/staff/b18346.

Watson, J. B. (1925). *Behaviorism*. New York: Norton.

Watson, J. D. (1972). *The Double Helix*. New York: Athenaeum.

Wellesley, K. (1995). *Tacitus, the Histories*. London: Penguin Books.

Westfall, R. (1993). *The Life of Isaac Newton*. Cambridge: Cambridge University Press.

Weatherly, E. H. (1962). *Renaissance*. New York: Dell Publishing.

"What Are We High on Now?" (September 8-15, 2014). *Time*, 66-67.

"What Makes Us Special" *Scientific American*, September 2014, Vol. 311(3), 60-61.

White, N. (1983). *Handbook of Epictetus*. Indianapolis, IN: Hackett Publishing Co.

Willougby, R. (1997). *Life in Medieval England*. Andover, UK: Pitkin.

Wills, G. (1978). *Inventing America*. New York: Doubleday.
—— (1980). *Explaining America, the Federalist*. New York: Doubleday.
Wilson, E. O. (1975). *Sociobiology, the new synthesis*. Cambridge, MA: Harvard University Press.
—— (1998). *Consilience, the Unity of Knowledge*. New York: Knopf.
—— (1998). "Back from Chaos." *Atlantic Monthly*, *281* (3), 42-62.
—— (1978). *On Human Nature*. Cambridge: Harvard University Press.
—— (2014). *A Window on Eternity, A Biologist's Walk Through Gorongosa National Park*. New York: Simon & Schuster.
Wines, M., "Climate Change Threatens to Strip the Identity of Glacier National Park," *The New York Times*, November 23, 2014, p. 17.
Wood, B., "Welcome to the Family," *Scientific American*, September 2014, Vol. *311*(3), 43-47.
Wood, M. (1985). *In Search of the Trojan War*. New York: New American Library.
—— (1994). *In Search of the Dark Ages*. London: Penguin Books.
Wooley, C. L. (1965). *The Sumerians*. New York: W. W. Norton.
Wright, R. (1994). *The Moral Animal*. New York: Pantheon.
Xenophon (no date). *The Education of Cyrus*. (H. G. Dakyns, trans.). London: J. M. Dent & Sons.
Zakaria, F. (2012). *The post-American world: Release 2.0*. New York: W. W. Norton.
Zhang, J. & J. & X. He (April 2005). "Significant Impact of Protein Dispensability on the Instantaneous Rate of Protein Evolution." *Mol. Biol. Evol.*, *22*, 1147–1155.
Zimmer, C. (2006). *Evolution, The Triumph of an Idea*. New York: Harper Perennial.
—— (2008). "What is a Species?" *Scientific American*, *298*(6), 72–79.

Index

A

B

C

F

G

H

I

J

N

O

P

Q

R

S

T

U

V